ELECTRONICS LABORATORY MANUAL

(Hands on Practice)

Ugochukwu Edebeani Anionovo

Department of Electrical Engineering

Copyright © 2024 Ugochukwu Edebeani Anionovo

All rights reserved

This Electronics Laboratory Manual: Hands on Practice, by Ugochukwu E. Anionovo is copyrighted under the terms of a Creative Commons license:

ISBN: 9798304843577

Cover design by: Author
Electrical and Power System Engineering
Department of Electrical Engineering,
Nnamdi Azikiwe University, Awka
Enugu-Onitsha Expressway,
Ifite Road, 420110, Awka.
ue.anionovo@unizik.edu.ng
Printed in Nigeria

This manual is dedicated to my Parents

CONTENTS

Title Page
Copyright
Dedication
Preface
Electronics Laboratory Manual 1
Safety Rules and Operating Procedures 3
Laboratory Safety Information 5
Trouble Shooting Hints 6
Chapter One 7
Chapter Two 15
Chapter Three 20
Chapter Four 25
Chapter Five 29
Chapter Six 33
Chapter Seven A 37
Chapter Seven B 42
Chapter Seven C 49
Chapter Eight 55
Chapter Nine 59
Chapter Ten 63
Chapter Eleven 67
Chapter Twelve 70
Chapter Thirteen 74
Chapter Fourteen 79
Chapter Fifteen 84
Chapter Sixteen 88
Chapter Seventeen 93
Chapter Eighteen 98

Chapter Nineteen	103
Chapter Twenty	109
Chapter Twenty-One	114
Chapter Twenty-Two	118
Chapter Twenty-Three	122
Chapter Twenty-Four	126
Chapter Twenty-Five	132
Chapter Twenty-Six	137
Chapter Twenty-Seven	142
Chapter Twenty-Eight	146
Chapter Twenty-Nine	149
Chapter Thirty	158
Appendix A: Creating Graphs Using a Spreadsheet	165
Appendix B: Manufacturer's Datasheet Links	167
Appendix C: Component Symbol Glossary	168
References	171

PREFACE

This laboratory manual is intended for use in fundamental electronics or introductory semiconductor devices courses and is appropriate for two and four year electrical engineering technology curriculums. The manual contains sufficient exercises for two 15 week courses using a two to three hour practicum period. It assumes familiarity with basic electrical circuit analysis techniques and theorems. The topics cover basic diodes through DC biasing and AC analysis of small signal bipolar and Field Effect Transistor (FET) amplifiers along with class A and B large signal analysis, Soldering and Soldering Technique. For equipment, each lab station should include a dual adjustable Direct Current (DC) power supply, a dual trace oscilloscope, a function generator and a quality Digital Multi-meter (DMM). Some exercises also make use of a distortion analyzer and a low distortion generator (generally, Total Harmonics Distortion -THD below 0.01%), although these portions may be bypassed. For components, a selection of standard value ¼ watt carbon film resistors ranging from a few ohms to a few mega ohms is required along with an array of typical capacitor values (film types recommended below 1μF and aluminum electrolytic above). Specialty passives include a Cadmium Sulphide (CdS) or a large area photo resistor cell, thermistor and a 20 ohm, 20 watt load resistor. A decade resistance box and a 10 kΩ potentiometer may also be useful. Active devices include small signal diodes such as the 1N914 or 1N4148, rectifying diodes such as the 1N4000 series, the NZX5V1B or 1N751 Zener, single LEDs of various colors, a super bright LED, 2N3904 or 2N2222 NPN transistor, 2N3906 PNP transistor, and MPF102 N channel Junction Field Effect Transistor (JFET). A small 12.6 Volts center tapped (VCT) power transformer is used in the power supply project and associated exercises along with a three-terminal linear regulator example IC 7805.

Each exercise begins with an Objective and a Theory Overview. The Equipment List follows with space provided for serial numbers and measured values of components. Schematics are presented next along with the step-by-step procedure. Many exercises include sections on troubleshooting and design. Simulations are often presented as well, and any quality simulation package such as LTspice, TINA-TI, Multisim, PSpice, or PSIM Professional can be used. All data tables are grouped together, typically with columns for the theoretical and experimental results, along with a column for the percent deviations between them. Finally, a group of appropriate questions are presented.

ELECTRONICS LABORATORY MANUAL
(Hands on Practice)

by
Ugochukwu Edebeani Anionovo

Version 1.0, 20th December 2024

This **Electronics Laboratory Manual: Hands on Practice**, by Ugochukwu E. Anionovo is copyrighted under the terms of a Creative Commons license:

This work is freely redistributable for non-commercial use, share-alike with attribution

Published by **Ugochukwu E. Anionovo** via Amazon KDP

ISBN: 9798304843577

For more information or feedback, contact:
Ugochukwu E. Anionovo (PhD),

UGOCHUKWU EDEBEANI ANIONOVO PHD

Electrical and Power System Engineering
Department of Electrical Engineering,
Nnamdi Azikiwe University, Awka
Enugu-Onitsha Expressway,
Ifite Road, 420110, Awka.
ue.anionovo@unizik.edu.ng

.

"It doesn't matter how beautiful your theory is, it doesn't matter how smart you are. If it doesn't agree with experiment, it's wrong."
- **Richard Feynman**

SAFETY RULES AND OPERATING PROCEDURES

Emergency Response
1. It is your responsibility to read safety and fire alarm posters and follow the instructions during an emergency.
2. Know the location of the fire extinguisher in your lab and know how to use them.
3. Know the location and function of all laboratory safety equipment.
4. Know the building evacuation procedures.
5. Note the location of the Emergency Disconnect (red button near the door) to shut off power in an emergency. Note the location of the nearest telephone (map on bulletin board).

General Safety Rules
6. Read all directions for an experiment and follow the directions exactly as they are written. If in doubt, ask the teacher.
7. Always obtain permission before experimenting on your own.
8. Never handle any equipment unless you have specific permission.
9. Report any accident or injury, no matter how small, to your teacher immediately.
10. Dispose of all material according to the teacher's instructions.
11. Never eat, drink, or smoke while working in the laboratory.
12. You have long hair or loose clothes, make sure it is tied back or confined.
13. Students are allowed in the laboratory only when the lab instructor is present.
14. Open drinks and food are not allowed near the lab benches.
15. Report any broken equipment or defective parts to the lab instructor. Do not open, remove the cover, or attempt to repair any equipment.
16. When the lab exercise is over, all instruments, except computers, must be turned off. Return substitution boxes to the designated location. Your lab grade will be affected if your laboratory station is not tidy when you leave.
17. University property must not be taken from the laboratory.
18. Do not move instruments from one lab station to another lab station.
19. Do not tamper with or remove security straps, locks, or other security devices. Do not disable or attempt to defeat the security camera.
20. Anyone violating any rules or regulations may be denies access to these facilities.

Electrical Safety
1. Obtain permission before operating any high voltage equipment.
2. Make sure electronic equipment is OFF when plugging or unplugging from an outlet.
3. Make sure the work area for electrical equipment is clean and dry.

4. Do not "daisy-chain" electrical power cords.
5. Maintain an unobstructed access to all electrical panels.
6. Avoid using extension cords whenever possible. If you must use one, obtain a heavy- duty one that is electrically grounded, with its own fuse, and install it safely. Extension cords should not go under doors, across aisles, be hung from the ceiling, or plugged into other extension cords.
7. Never, ever modify, attach or otherwise change any high voltage equipment.
8. Always make sure all capacitors are discharged (using a grounded cable with an insulating handle) before touching high voltage leads or the "inside" of any equipment even after it has been turned off. Capacitors can hold charge for many hours after the equipment has been turned off.

End of Laboratory activity rules

1. Clean all laboratory equipment and return to their locations
2. Unplug and store properly any electrical device.
3. Wash your hands after every experiment.
4. Clean up your work area before leaving.

Certification

I have read and understand these rules and procedures. I agree to abide by these rules and procedures at all times while using these facilities. I understand that failure to follow these rules and procedures will result in my immediate dismissal from the laboratory and additional disciplinary action may be taken.

Signature: _____   Date: _____   Lab. No.: _____

LABORATORY SAFETY INFORMATION

The danger of injury or death from electrical shock, fire, or explosion is present while conducting experiments in this laboratory. To work safely, it is important that you understand the prudent practices necessary to minimize the risks and what to do if there is an accident.

Electrical Shock:
Avoid contact with conductors in energized electrical circuits. Electrocution has been reported at dc voltages as low as 42 volts. Just 100 mA of current passing through the chest is usually fatal. Muscle contractions can prevent the person from moving away while being electrocuted.

Do not touch someone who is being shocked while still in contact with the electrical conductor or you may also be electrocuted. Instead, press the Emergency Disconnect (red button located near the door to the laboratory). This shuts off all power, except the lights.

Make sure your hands are dry. The resistance of dry, unbroken skin is relatively high and thus reduces the risk of shock. Skin that is broken, wet or damp with sweat has a low resistance.

When working with an energized circuit, work with only your right hand, keeping your left hand away from all conductive material. This reduces the likelihood of an accident that results in current passing through your heart.

Be cautious of rings, watches, and necklaces. Skin beneath a ring or watch is damp, lowering the skin resistance. Shoes covering the feet are much safer than sandals.

If the victim isn't breathing, find someone certified in CPR. Be quick! Some of the staff in the Department Offices are certified in CPR. If the victim is unconscious or needs an ambulance, contact the Department Office for help or call the state emergency number. If able, the victim should go to the Student Health Services for examination and treatment.

Fire:
Transistors and other components can become extremely hot and cause severe burns if touched.

If resistors or other components on your proto-board catch fire, turn off the power supply and notify the instructor. If electronic instruments catch fire, press the Emergency Disconnect (red button). These small electrical fires extinguish quickly after the power is shut off. Avoid using fire extinguishers on electronic instruments.

Explosion:
When using electrolytic capacitors, be careful to observe proper polarity and do not exceed the voltage rating. Electrolytic capacitors can explode and cause injury. A first aid kit is located on the wall near the door. Proceed to Student Health Services, if needed.

TROUBLE SHOOTING HINTS

1. Be sure that the power is turned on.
2. Be sure the ground connections are common.
3. Be sure the circuit you built is identical to that in the diagram. (Do a node-by-node check)
4. Be sure that the supply voltages are correct.
5. Be sure you plug in cable to the right terminal in the multi-meter to measure the voltage/resistance (upper terminal) or the current (lower terminal).
6. Be sure that the equipment is set up correctly and you are measuring the correct parameter.
7. Be sure the BJT's collector and emitter terminals are in correct orientation.
8. If steps 1 through 5 are correct, then you probably have used a component with the wrong value or one that doesn't work. It is also possible that the equipment does not work (although this is not probable) or the protoboard you are using may have some unwanted paths between nodes.

To find your problem you must trace through the voltages in your circuit node by node and compare the signal you have to the signal you expect to have. Then if they are different use your engineering judgment to decide what is causing the different or ask your lab assistant.

CHAPTER ONE
Introduction to Electronics Lab

1.1 Objective

The laboratory emphasizes the practical, hands-on component of this course. It complements the theoretical material presented in lecture, and as such, is integral and indispensible to the mastery of the subject. There are several items of importance here including proper safety procedures, required tools, and laboratory reports. This exercise will finish with a section on component identification.

1.2 Lab Safety and Tools

If proper procedures are followed, the electrical lab is a perfectly safe place to work. There are some basic rules: No food or drink is allowed in lab at any time. Liquids are of particular danger as they are ordinarily conductive. While the circuitry used in lab normally presents no shock hazard, some of the test equipment may have very high internal voltages that could be lethal (in excess of 10,000 volts) along with the 120 V Alternating Current (AC) power used to operate the equipment that can also be lethal if good safety practices are not followed. Spilling a bottle of water or soda onto such equipment could leave the experimenter in the receiving end of a severe shock. Similarly, items such as books and jackets should not be left on top of the test equipment as it could cause overheating. Use caution in storing these items during lab periods to avoid trip or fall hazards in the lab.

Each lab bench is self contained. All test equipments are arrayed along the top shelf. Built into the bench is a power strip. All test equipment for this bench should be plugged into this strip. None of this equipment should be plugged into any other strip. This strip is controlled by a circuit breaker. In the event of an emergency, all test equipment may be powered off through this one switch. Further, the benches are controlled by dedicated circuit breakers in the main lab panel. Located at the front of the lab is an A/B/C class fire extinguisher suitable for electrical fires. Also at the front of the lab is a safety kit. This contains bandages, cleaning swaps and the like for small cuts and the like. Familiarize yourself with the location of these items in the lab. For serious injury, the Campus or factory Security Office should be contacted.

A lab bench should always be left in a secure mode. This means that the power to each piece of test equipment should be turned off; the bench itself should be turned off, all AC and DC power and signal sources should be turned down to zero, and all other equipment and components properly stowed with lab stools pushed under the bench. Any cables or cords used in the lab should be stored properly after the exercise is completed.

It is important to come prepared to lab. This includes the class text, the lab exercise for that day, class notebook, calculator, and hand tools. The tools include an electronic breadboard, test leads, wire-strippers, and needle nose pliers or hemostats. A small pencil soldering iron may also be useful. A basic digital multi-meter (DMM) rounds out the list. A typical breadboard or protoboard is shown in Plate 1.1.

Plate 1.1: Sample Breadboard   Plate 1.2: Sample Breadboard

This particular item features two main wiring sections with a common strip section down the center.
Boards can be larger or smaller than this and may or may not have the mounting plate as shown. The connections are spaced 0.1 inch apart which is the standard spacing for many semiconductor chips. These are clustered in groups of five common terminals to allow multiple connections. The exception is the common strip which may have dozens of connection points. These are called buses and are designed for power and ground connections. Interconnections are normally made using small diameter solid hookup wire, usually AWG 22 or 24. Larger gauges may damage the board while smaller gauges do not always make good connections and are easy to break.
In the plate 1.2, the color highlighted sections indicate common connection points. Note the long blue section which is a bus. This unit has four discrete buses available. When building circuits on a breadboard, it is important to keep the interconnecting wires short and the layout as neat as possible. This will aid both circuit functioning and ease of troubleshooting.

1.3 Component Identification

In this lab, many different electronic components are used including passive devices such as resistors and capacitors as well as semiconductors such as diodes and transistors, and finally, integrated circuits. These devices are available in many different case styles. Two broad classifications are **through-hole and surface mount.** As circuits will be built on protoboards or breadboard, through-hole components are of particular interest here.
Surface mount devices are generally smaller and use thin flat tabs or stubs in place of ordinary wire leads.
In production they are soldered directly to the surface of the printed circuit board without the use of holes.

1.3.1 Resistors and Potentiometers

ELECTRONICS LABORATORY MANUAL

Resistors are perhaps the single most common component; plate 1.3a. They are classified as passive devices (versus active devices aka semiconductors). Resistors have two leads and are not directional so they cannot be inserted backwards. Leads are usually axial (i.e., emanating from opposite ends). The physical size of a resistor indicates its power handling capacity, not its resistance. The general purpose lab resistor is usually a carbon film type, ¼ watt dissipation. Resistance values are shown via a color coded series of bands for most types, although high precision resistors may have the value printed directly on the body.

Potentiometers may be either rotary or linear travel (slider), with rotary being the most common; plate 1.3a. Most rotary pots are ¾ turn, although precision trim pots may be 20 turns or more. Typically, the center of the three connections is the wiper arm. Rotary pots may be designed for panel mount (for example, a volume control on a stereo) or board mounts (such as a calibration control). The taper of a pot indicates how resistance and position are related. Pots may have a linear taper or a specialized audio taper (log taper). A linear taper means that a specific degree of rotation will produce the same resistance change. Rotating the shaft half way, for example, results in a 50/50 split of the resistance. In contrast, an audio taper pot would show a 10/90 split. Pots are also available in multi-gang, that is, several pots controlled by one common shaft.

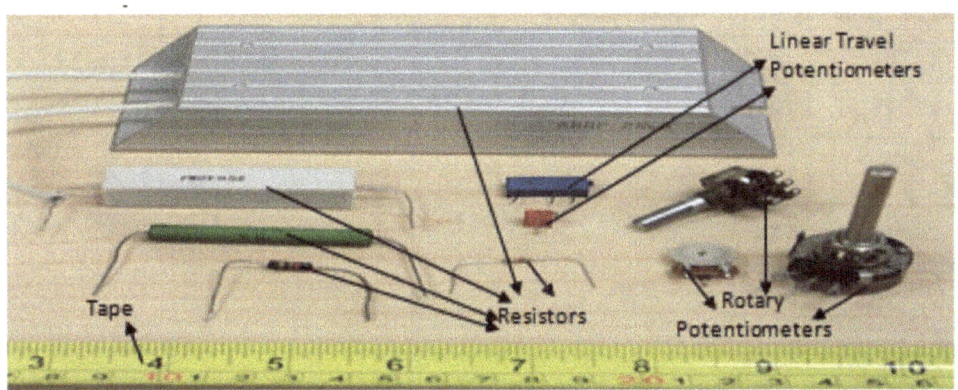

Plate 1.3a: Resistors and Potentiometers.

Resistor Color Coding
Resistors are coded with colors to identify their value. Plate 1.3b shows Resistor Color Codes Chart.

Mnemonics
Useful mnemonics have been created to make it easier to remember the numeric order of resistor color bands: The following example includes the tolerance codes — gold, silver and none:

Bad **B**eer **R**ots **O**ut **Y**our **G**uts **B**ut **V**odka **G**oes **W**ell – **G**et **S**ome **N**ow.

The colors are sorted in ascending order of frequency to make them easy to remember and to reduce the significance of possible read errors due to color shifts and fading over time: **R**ed (2), **O**range (3), **Y**ellow (4), **G**reen (5), **B**lue (6), and **V**iolet (7). **B**lack (0) has no energy, **B**rown (1) has a little more, **W**hite (9) has everything and **G**rey (8) is like white, but less intense

Color	Signficant figures			Multiply	Tolerance (%)	Temp. Coeff. (ppm/K)	Fail Rate (%)
black	0	0	0	× 1		250 (U)	
brown	1	1	1	× 10	1 (F)	100 (S)	1
red	2	2	2	× 100	2 (G)	50 (R)	0.1
orange	3	3	3	× 1K		15 (P)	0.01
yellow	4	4	4	× 10K		25 (Q)	0.001
green	5	5	5	× 100K	0.5 (D)	20 (Z)	
blue	6	6	6	× 1M	0.25 (C)	10 (Z)	
violet	7	7	7	× 10M	0.1 (B)	5 (M)	
grey	8	8	8	× 100M	0.05 (A)	1 (K)	
white	9	9	9	× 1G			
gold			3th digit only for 5 and 6 bands	× 0.1	5 (J)		
silver				× 0.01	10 (K)		
none					20 (M)		

6 band — 3.21kΩ 1% 50ppm/K

5 band — 521Ω 1%

4 band — 82kΩ 5%

3 band — 330Ω 20%

Gap between band 3 and 4

indicates reading direction

Plate 1.3b: Resistor color coding chart

1.3.2 Capacitors

Capacitors are also classified as passives or passive devices, and can be thought of as very short term energy storage devices; plate 1.4.

Capacitors are dual lead but may have either axial or radial (radiating from one end) leads. Unlike resistors, the physical size of a capacitor offers a clue as to its capacitance and voltage rating. All other factors being equal the greater the capacitance or voltage rating, the larger the capacitor. Smaller capacitors (below 1 µF) are not polarized and can be inserted into a circuit either way. The more popular dielectrics for this range include the ceramics (usually disk or coin shaped) and poly film types (polyester, polypropylene, etc.) which are usually block shaped. Teardrop shaped tantalum capacitors are used commonly for power supply bypass. They are polarized and must be inserted in the circuit in the specified direction. Larger capacitance values (over 1 µF) are often realized via aluminum electrolytic. These are also polarized. Failure to insert these in the proper direction may result in unpredictable results, including the capacitor exploding. While they do not perform as well as film types in terms of leakage, accuracy,

etc., they are offer high volumetric efficiency (i.e., small physical size given the capacitance). Very large caps may have screw terminals in place of wire leads. In years past, color coding was common but this has generally been replaced with values printed directly on the body of the capacitor. Sometimes a numeric code is used such as "102". This is read as 10 followed by 2 zeroes, with the result in picofarads, or 1000 pF (1 nF) in this case. Finally, because capacitors are charge storage devices, they may present a shock hazard from stored charge after they are removed from a circuit. This charge may be bled off with a low value resistance placed across the leads.

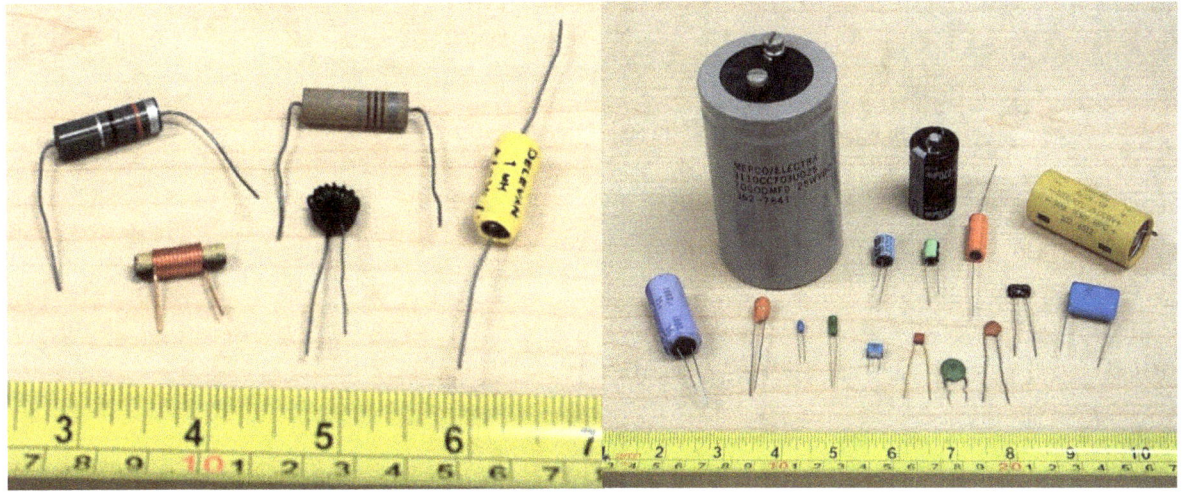

Plate 1.4: Capacitors   Plate 1.5: Inductors

1.3.3 Inductors

The third and final passive device is the inductor; plate 1.5. Also non-polarized, they normally have axial leads. Smaller values may be completely encapsulated and appear not much different from a carbon composition resistor. Others may use some form of jacket or coating while still others show bare wire (the wire only appears bare, it is in fact covered by a thin clear insulating coating). These vary from the size of small resistors to what appear to be large spools of wire.

1.3.4 Diodes

Diodes are a two lead semiconductor; plate 1.6. They are polarized and typically have axial leads. The two leads are referred to as the anode and cathode. Signal diodes are around the size of ¼ watt resistors and sometimes use a glass body. The cathode is marked by a band or stripe on the body of the diode. The cathode of an Light Emitting Diode (LED) is usually marked by a flat spot on the plastic housing or by the shorter of the two leads. High power diodes are much more robust and might appear at first glance to be a short bolt or stud with leads attached to it. Component numbers are usually stamped on the body of the device.

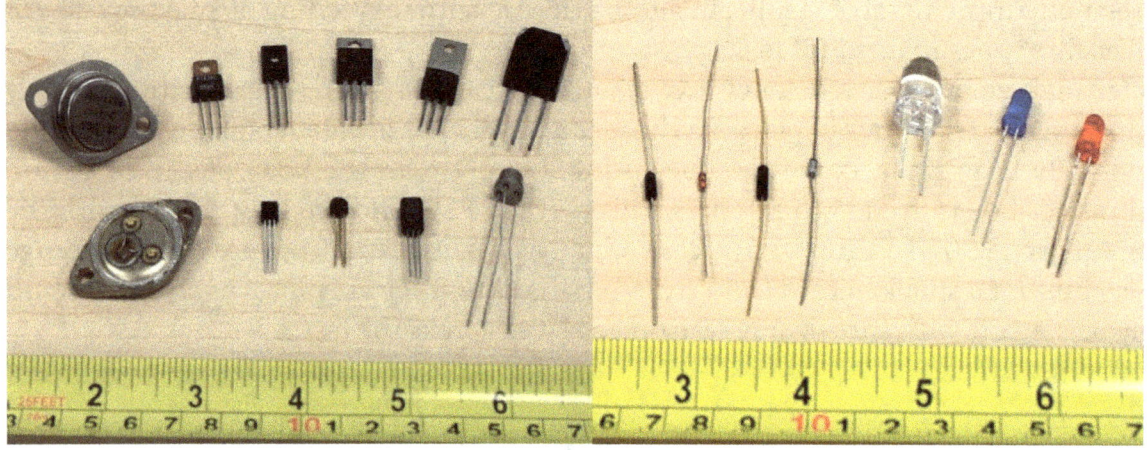

Plate 1.6: Diodes   Plate 1.7: Transistors

1.3.5 Transistors

There are many types of transistors; plate 1.7. Generally, they are three lead devices. Component model numbers will be stamped directly onto the case. Small power dissipation (< 500 mW) units will usually be seen in plastic TO-92 cases, round metal TO-5 cans or variations on the theme. Mid power devices typically use TO-220, TO-202 or the like "power tab" cases. For higher powers the oval TO-3 cases are employed. A similarly shaped but slightly smaller variant is the TO-66. Power devices will need to use a heat sink to keep them cool. TO-92 cases use a flattened front face so that the three pins may be distinguished from each other without confusion. The round TO-5 can uses a small tab to indicate pin 1.

1.3.6 Integrated Circuits

There are a very wide variety of integrated circuits; plate 1.8. Multi-lead versions of the TO-5 can are sometimes used but the most common through-hole package is the Dual In-Line Package, denoted as DIP or DIL. A single in-line package is also available for some functions. High power devices often use multi-lead versions of the popular TO-220 and TO-3 case styles. Like other semiconductors, component model numbers are printed directly on the package. A notch or dimple will denote which lead is pin 1 on the DIP/DIL cases. Plate 1.9 shows different electronic package style for transistors as we as integrated circuits and the "TO" stands for transistor outline.

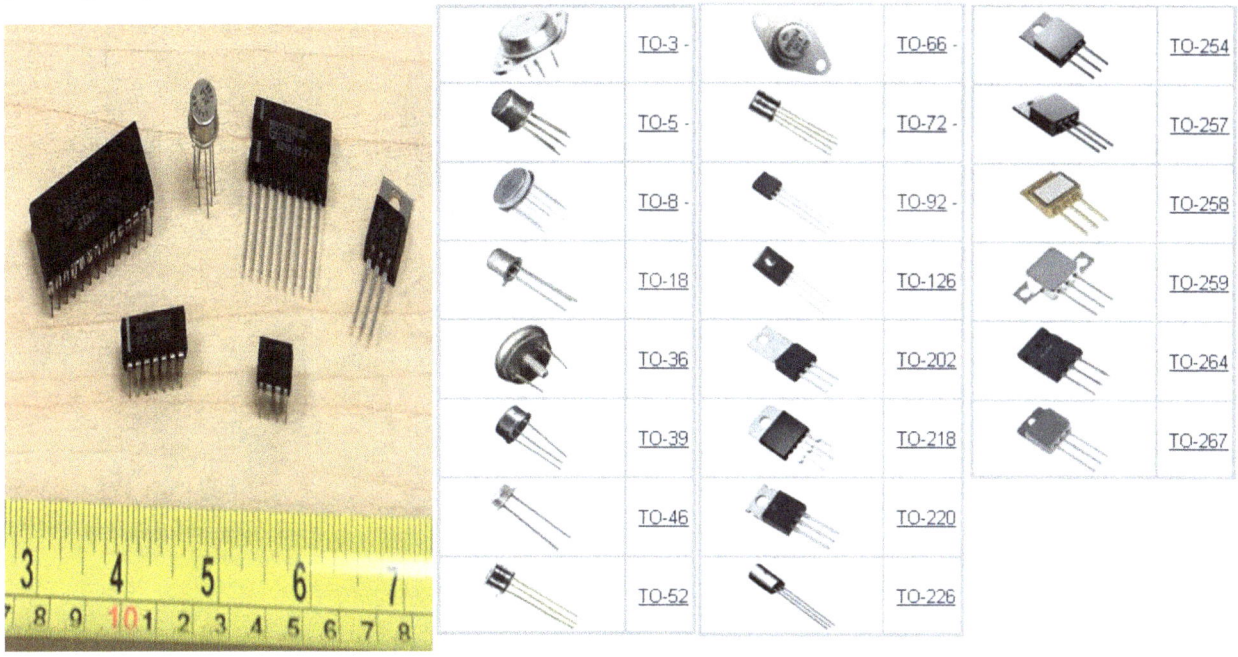

Plate 1.8: Integrated Circuits   Plate 1.9: TO Package styles

1.3.7 Transformers

Transformers can vary from tiny audio devices to room size devices used in power generation and distribution; plate 1.10. No matter the size, their job is a simple one: to isolate the source and load, to match two different impedance devices or to change the voltage level. A very common application is stepping down a 120 Volt AC line voltage to a more modest level so that it can be rectified, filtered, and turned into a stable DC source to drive electronic circuits. Besides the voltage turns ratio, the most important characteristic is the VA or volt-amps rating of the device. All other factors being equal, the higher the VA rating, the larger the transformer. Transformers applicable for consumer electronics may be either chassis mount with leads or Printed Circuit Board (PCB) mount with through-hole pins. Transformers only operate with AC voltages.

Plate 1.10: Transformers   Plate 1.11: Heat sinks

1.3.8 Heat Sinks

Heat sinks are not a device, per se, but they are essential tools of semiconductor heat management. Their job is to effectively move heat from the semiconductor's case to the surrounding air, keeping the semiconductor cool. They range in size from small clip-ons to large extruded aluminum finned plates; plate 1.11. Some cases, such as the body of the TO-3 or the tab of the TO-220 are electrically live. To prevent possible shorts and a live chassis, non-conductive isolating tabs and grommets are used to attach the semiconductors to the heat sink.

CHAPTER TWO
Resistive Sensors

2.1 Objective

The objective of this exercise is to investigate devices that can be used to sense environmental factors such as light and temperature. These are important if circuitry is to react to surrounding conditions, for example, controlling fan speed that is proportional to temperature or turning lights on or off depending on existing light levels. Two such devices are the light dependent resistor or LDR, and the thermistor or temperature dependent resistor. They can be thought of as resistors whose values depend on either the surrounding light levels or the temperature.

2.2 Theory Overview

One typical LDR is the CdS (Cadmium Sulfide) cell. The resistance of the CdS cell is inversely proportional to light levels. In darkness, it may exhibit a resistance of tens or even hundreds of kilo ohms.

Under high brightness, the resistance may be as little as a few hundred ohms. Thermistors come in two types: PTC or Positive Temperature Coefficient whose resistive value increases with temperature, and

NTC or Negative Temperature Coefficient whose resistance value decreases with increasing temperature. In contrast, ordinary resistors are designed to be immune to temperature change as much as possible.

One way of using these devices is by placing them in a voltage divider. The resulting voltage will reflect the light levels or temperature. Depending on the position of the device, the voltage can be made to either increase or decrease as the environmental factor changes. For example, the voltage could rise as temperature rises but it could also be designed to have the voltage decrease as temperature rises. Both functions have their uses. Finally, it is worth noting that these devices do not necessarily respond immediately to environmental changes. For example, a thermistor might be used to sense air temperature.

If the air temperature were to suddenly rise, there would be some time lag in the response of the thermistor. This is due to the fact that the thermistor itself has mass and requires some time to either heat up or cool down.

2.3 Experimental Example
2.3.1 Equipment

(1) Adjustable DC power supply model:_____ S/NO.:_____
(1) DMM model:_____ S/NO.:_____
(1) Non-diffuse light source (pen light)
(1) Heat source (diffused light duty heat gun or blow dryer)
(1) Nominal 1 kΩ – 10 kΩ CdS cell (GL5528)
(1) 10 kΩ @ 25°C NTC thermistor (Vishay NTCLE100E3)

(1) 10 k Ω resistor ¼ watt actual: _____

CdS cell datasheet: http://cdn.sparkfun.com/datasheets/Sensors/LightImaging/SEN-09088.pdf
Thermistor datasheet: http://www.vishay.com/docs/29049/ntcle100.pdf

Plate 2.1: Pen light   Plate 2.2: Heat gun, and Hair dryer or blow dryer.

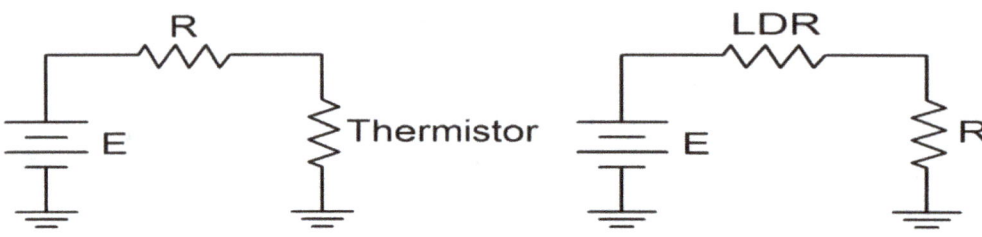

Figure 2.1: LDR test schematics   Figure 2.2: Thermistor test schematics

2.3.2 Procedure
LDR
1. Use the DMM to measure the resistance of the LDR. Do this away from any windows and do not block ambient room lighting from hitting the device. Record the resistance value in Table 2.1 under "Normal".
2. Place a finger over the LDR to block all light, making sure that the leads are not also touched. Record the resulting resistance value under "Dark".
3. Shine the pen light directly onto the LDR at a distance of about 25 centimeters. Record the result under "Bright".
4. Construct the circuit of Figure 2.1 using E=10 volts and R=10 k Ω. Using the voltage divider rule, determine the expected value for the voltage across R under normal lighting and then measure the voltage. Record these values in Table 2.2.
5. Repeat step 4 for the Dark and Bright conditions.
6. Finally, slowly move the pen light toward and away from the LDR. Note what happens to the voltage, recording the maximum and minimum voltages obtained in Table 2.3.

Thermistor

7. Use the DMM to measure the resistance of the LDR at room temperature and record the result in Table 2.4. Do not handle the device excessively as body heat may affect it.

8. Build the circuit of Figure 2.2 using E=10 volts and R=10 kΩ. Measure the voltage across the thermistor and record it in Table 2.5 under "Room Temp".
9. Monitor the thermistor voltage while applying heat. Caution: If you are using a standard heat gun, place it on a low setting, use a diffuser or keep the gun at least a half meter away to avoid possibly damaging connecting wires or the protoboard. After 30 to 60 seconds, record the thermistor voltage in Table 2.5. Turn off the heat source and note how long it takes the thermistor circuit to recover back to the original reading.

2.3.3 Result Tables

Variation	V_R Theory	V_R Experiment	% Deviation
Normal			
Dark			
Bright			

Table 2.1

Variation	Resistance
Normal	
Dark	
Bright	

Table 2.2

Table 2.3

Variation	
Maximum	
Minimum	

Table 2.4

Variation	$V_{thermistor}$
Room Temp	
Hot	

Table 2.5

R at Room Temperature	
V_R	

2.3.4 Questions

1. If the LDR and resistor positions had been swapped, how would the values of Table 2.2 change?

_____
_____
_____
_____
_____

2. Would the voltages measured in Table 2.2 change appreciably if R had been 1 kΩ instead of 10 kΩ?

_____
_____
_____
_____

3. If the positions of the thermistor and resistor in Figure 2.2 had been swapped, how would the Table 2.5 values change?

4. If 20 volt power sources had been used, how would the values of Tables 2.2, and 2.5 change, if at all?

Extra: Voltage divider

In electronics, a voltage divider (also known as a potential divider) is a passive linear circuit that produces an output voltage (V_{out}) that is a fraction of its input voltage (V_{in}). Voltage division is the result of distributing the input voltage among the components of the divider. A simple example of a voltage divider is two resistors connected in series, with the input voltage applied across the resistor pair and the output voltage emerging from the connection between them.

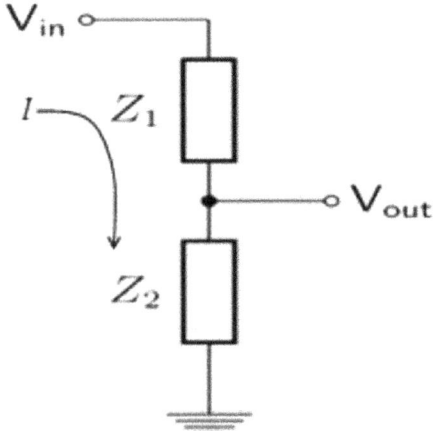

Figure 2.3: voltage divider schematic

A voltage divider referenced to ground is created by connecting two electrical impedances in series, as shown in Figure 2.3. The input voltage is applied across the series impedances Z_1 and Z_2 and the output is the voltage across Z_2. Z_1 and Z_2 may be composed of any combination of elements such as resistors, inductors and capacitors. If the current in the output wire is zero then the relationship between the input voltage, V_{in}, and the output voltage, V_{out}, is mathematically evaluated as shown in equation (2.1)

$$V_{out} = \frac{Z_2}{Z_1 + Z_2} \cdot V_{in} \qquad (2.1)$$

Using Ohm's law, equation (2.1) can be proven as follows:

$$V_{in} = I.(Z_1 + Z_2) \qquad (2.2)$$

$$V_{out} = I.Z_2 \qquad (2.3)$$

$$I = \frac{V_{in}}{Z_1 + Z_2} \text{ (From equation 2.2)} \qquad (2.4)$$

$$V_{out} = V_{in} \cdot \frac{Z_2}{Z_1 + Z_2} \text{ (Substituting equation 2.4 into 2.3)} \qquad (2.5)$$

CHAPTER THREE
Diode Curves

3.1 Objective
The objective of this exercise is to examine the operation of the basic switching diode and to plot its characteristic curve. Basic DC circuit operation will also be examined.

3.2 Theory Overview
The basic diode is an asymmetric non-linear device. That is, its current-voltage characteristic is not a straight line and it is sensitive to the polarity of an applied voltage or current. When placed in forward bias (i.e. positive polarity from anode to cathode), the diode will behave much like a shorted switch and allow current flow. When reversed biased the diode will behave much like an open switch, allowing little current flow. Unlike a switch, a silicon diode will exhibit an approximate 0.7 volt drop when forward biased. The precise voltage value will depend on the semiconductor material used. This volt drop is sometimes referred to as the knee voltage as the resulting I-V curve looks something like a bent knee.

The effective instantaneous resistance of the diode above the turn-on threshold is very small, perhaps a few ohms or less, and is often ignored. Analysis of diode circuits typically proceeds by determining if the diode is forward or reversed biased, substituting the appropriate approximation for the device, and then solving for desired circuit parameters using typical analysis techniques. For example, when forward biased, a silicon diode can be thought of as a fixed 0.7 volt drop, and then Kirchhoff's Voltage Law (KVL) and Kirchhoff's Current Law (KCL) can be applied as needed. The polarity of the device is typically denoted by a band placed closest to the cathode.

3.3 Experimental Example
3.3.1 Equipment
(1) Adjustable DC power supply model:_____ S/NO.:_____
(1) DMM model:_____ S/NO.:_____
(2) Signal diodes (1N4148, 1N914)
(1) 1 k Ω resistor ¼ watt actual: _____
(1) 10 k Ω resistor ¼ watt actual: _____
(1) 4.7 k Ω resistor ¼ watt actual: _____
1N4148/1N914 Datasheet: https://www.onsemi.com/pub/Collateral/1N914-D.PDF

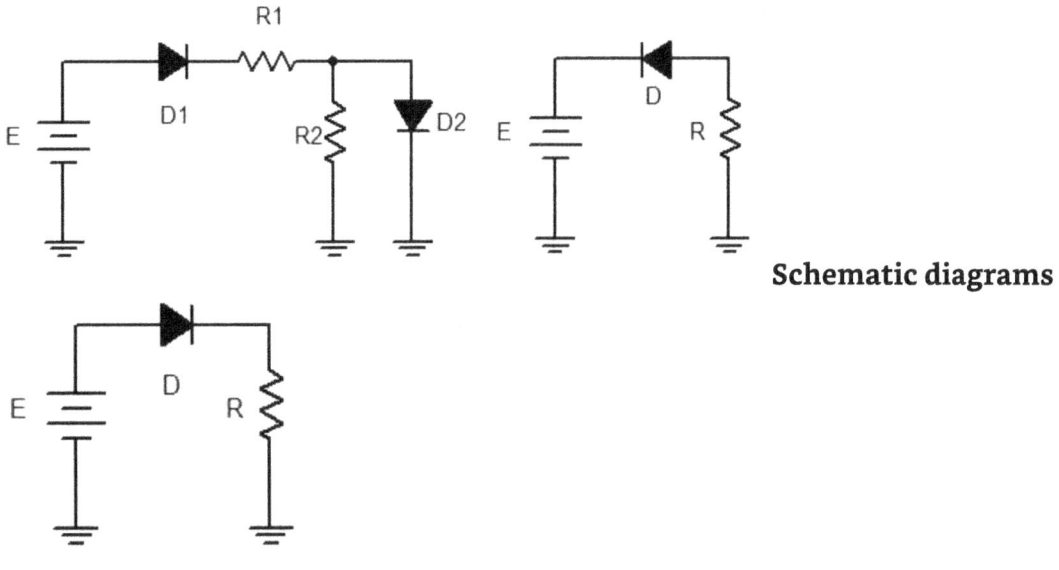

Schematic diagrams

Figure 3.1   Figure 3.2   Figures 3.3

3.3.2 Procedure
Forward Curve
1. Consider the circuit of Figure 3.1 using R = 1 kΩ. For any positive value of E, the diode should be forward biased. Once E exceeds the knee voltage, all of E (minus approximately 0.7 volts) drops across R. Thus, as E increases, so does the diode current.
2. Build the circuit of Figure 3.1 using R = 1 kΩ. Set E to 0 volts and measure both the diode's voltage and current and record the results in Table 3.1. Remember, voltage is measured across a device (parallel) while current is measured through it (series). Repeat this process for the remaining source voltages listed.
3. From the data collected in Table 3.1, plot the current versus voltage characteristic of the forward biased diode. Make sure VD is the horizontal axis with ID on the vertical.

Reverse Curve
4. Consider the circuit of Figure 3.2 using R = 1 kΩ. For any positive value of E, the diode should be reversed biased. In this case, the diode should always behave like an open switch and thus no current should flow. If no current flows, the voltage across R should be zero, and thus the diode voltage should be equal to the applied source voltage. Note that the diode's voltage polarity is negative with respect to that of Figure 3.1.
5. Build the circuit of Figure 3.2 using R = 1 kΩ. Set E to 0 volts and measure both the diode's voltage and current and record the results in Table 3.2. Repeat this process for the remaining source voltages listed.
6. From the data collected in Table 3.2, plot the current versus voltage characteristic of the reverse biased diode. Make sure VD is the horizontal axis with ID on the vertical.

Practical Analysis
7. Consider the circuit of Figure 3.3 using E = 12 volts, R1 = 10 kΩ and R2 = 4.7 kΩ.

Analyze the circuit using the ideal 0.7 volt forward drop approximation and determine the voltages across the two resistors. Record the results in the first two columns of the first row (Variation 1) of Table 3.3.

8. Build the circuit of Figure 3.3 using E = 12 volts, R1 = 10 kΩ and R2 = 4.7 kΩ. Measure the voltages across the two resistors. Record the results in columns three and four of the first row (Variation 1) of Table 3.3. Also compute and record the percent deviations in columns four and five.
9. Reverse the direction of D1 and repeat steps 7 and 8 as Variation 2 in Table 3.3.
10. Return D1 to the original orientation and reverse the direction of D2. Repeat steps 7 and 8 as Variation 3 in Table 3.3.
11. Reverse the direction of both D1 and D2, and repeat steps 7 and 8 as Variation 4 in Table 3.3.

Computer Simulation

12. Repeat steps 7 through 11 using a simulator, recording the results in Table 3.4.

3.3.3 Result Tables

E(volts)	V_D	I_D
0		
1		
2		
5		
10		
15		

Table 3.1   Table 3.2

E(volts)	V_D	I_D
0		0
0.5		
1		
2		
4		
6		
8		
10		

Table 3.3

Variation	$V_{R1Theory}$	$V_{R2Theory}$	$V_{R1Exp.}$	$V_{R2Exp.}$	% Dev. V_{R1}	% Dev. V_{R2}
1						
2						
3						

	4						

Table 3.4

Variation	$V_{R1\,Sim}$	$V_{R2\,Sim}$
1		
2		
3		
4		

3.3.4 Questions

1. Is 0.7 volts a reasonable approximation for a forward bias potential? Is an open circuit a reasonable approximation for a reverse biased diode? Support your arguments with experimental data.

2. The "average" resistance of a forward biased diode can be computed by simply dividing the diode's voltage by its current. Using Table 3.1, determine the smallest average diode resistance (show work).

3. The instantaneous resistance (also known as AC resistance) of a diode may be approximated by taking the differences between adjacent current-voltage readings. That is, $r_{diode} = \Delta V_{diode}/\Delta I_{diode}$. What are the smallest and largest resistances using Table 3.1 (show work)? Based on this, what would a plot of instantaneous diode resistance versus diode current look like?

4. If the circuit of Figure 3.3 had been constructed with LEDs in place of switching diodes, would there be any changes to the values measured in Table 3? Why/why not?

CHAPTER FOUR
Light Emitting Diodes

4.1 Objective
This exercise examines the general performance and use of light emitting diodes. This includes forward bias and reverse bias characterization along with brightness variation.

4.2 Theory Overview
The LED is similar to the ordinary signal or rectifying diode in that it is polarity sensitive. In reverse bias the device behaves as an open and prevents current flow. In forward bias, the device allows current flow once its forward barrier potential is reached. This potential is significantly higher than that of ordinary diodes and depends on the material used, and hence, the color that is displayed. Generally, luminous intensity is a function of the forward current. That is, the greater the current, the brighter the output. In operation, a series limiting resistor or other control device must be used to limit the forward current and prevent damage that could occur to the LED from excessive current. Different technologies are used in the design and production of LEDs and there are many variations including full spectrum (white) and high brightness versions. The cathode of an LED is typically denoted by a flat spot on the plastic casing and/or by the shorter of the two leads.

4.3 Experimental Example
4.3.1 Equipment
(1) Adjustable DC power supply model:_____ S/No:_____
(1) DMM model:_____ S/No.:_____
(1) Each of standard LEDs of various colors (red, blue, green, yellow)
(1) High brightness white LED
(1) 1 k Ω resistor ¼ watt actual: _____

Standard red LED Datasheet:
https://www.sparkfun.com/datasheets/Components/LED/COM-09590-YSL-R531R3D-D2.pdf
High brightness white LED Datasheet:
http://cdn.sparkfun.com/datasheets/Components/General/YSL-R1042WC-D15.pdf

Schematic Diagrams

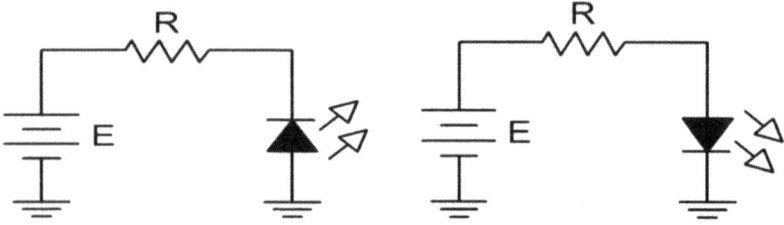

Figure 4.1   Figure 4.2

4.3.2 Procedure
Forward Curve
1. Consider the circuit of Figure 4.1 using R = 1 kΩ. For any positive value of E, the diode should be forward biased. Once E exceeds the knee voltage, the difference between the source and the knee drops across R. Thus, as E increases, so does the LED current and hence its brightness.
2. Build the circuit of Figure 4.1 using R = 1 kΩ and the red LED. Set E to 0 volts and measure both the LED voltage and current and record the results in Table 4.1. Note the relative brightness level. Repeat this process for the remaining source voltages listed.
3. From the data collected in Table 4.1, plot the current versus voltage characteristic of the forward biased LED. Make sure V_D is the horizontal axis with I_D on the vertical.
4. Repeat steps 2 and 3 for the blue LED using Table 4.2.
5. If other colors are available repeat steps 2 and 3 for them using Table 4.3 (create other tables as needed).

High Brightness
6. Replace the LED of Figure 1 with the high brightness white LED. Set the supply to 12 volts. Record the LED voltage, current and brightness in Table 4.4.

Reverse Curve
7. Consider the circuit of Figure 4.2 using R = 1 kΩ. For any positive value of E, the LED should be reversed biased. In this case, the LED should always be open causing no current to flow. If no current flows, the LED produces no light. Also, the voltage across R should be zero, and thus the LED voltage should be equal to the applied source voltage. Note that the LED voltage polarity is negative with respect to that of Figure 4.1.
8. Build the circuit of Figure 2 using R = 1 kΩ using the red LED. Set E to 0 volts and measure both the LED voltage and current and record the results in Table 5. Repeat this process for the remaining source voltages listed.
9. From the data collected in Table 5, plot the current versus voltage characteristic of the reverse biased diode. Make sure V_D is the horizontal axis with I_D on the vertical.

4.3.3 Result Tables
Table 4.1, Color: RED   Table 4.2 Color: Blue

E(volts)	V_D	I_D		Brightness

0			
1			
2			
3			
4			
6			
12			

E(volts)	V_D	I_D	Brightness
12			

Table 4.3, Color: Table 4.4 High Brightness

E(volts)	V_D	I_D	Brightness
0			
1			
2			
3			
4			
6			
12			

Table 4.5

E(volts)	V_D	I_D
0		
1		
2		
3		

4.3.4 Questions

1. Is the forward knee voltage of an LED comparable to that of ordinary switching and rectifying diodes?

   _____
   _____
   _____
   _____
   _____

2. Are the knee voltages of LEDs consistent across colors?

   _____
   _____

3. Compare the reverse characteristics of LEDs and switching diodes.

4. What can be said regarding LED brightness and current?

CHAPTER FIVE
Photodiodes

5.1 Objective
The objective of this exercise is to examine the operation of the photodiode in both the photovoltaic and photoconductive modes.

5.2 Theory Overview
The photodiode is, in essence, the reverse of the LED. In fact, depending on their design, LEDs can be used as a type of photodiode. Photodiodes are responsive to light in one of two ways. The first method is the photovoltaic mode. In this mode, a voltage appears across the PN junction that is proportional to the amount of light striking it. It can be thought of as a small voltage source or battery. The second mode is photoconductive. In this mode, the photodiode is reverse biased by an external DC supply. The amount of current flowing through the diode will be proportional to the amount of light striking the junction.
Typically, this current will pass through a series resistor to create a voltage or it can be sent into a current amplifier circuit.
A photo emitter/detector pair is a pairing of an LED and a photodiode that are designed to produce and detect the same wavelength of light. The wavelength of light may be outside the range of the human visible spectrum. Infrared (IR) is often used for consumer remote control devices. Emitter/detector pairs might use a phototransistor in place of a photodiode. The performance is similar except that photodiodes tend to have a quicker response while phototransistors tend to produce higher currents.

5.3 Experimental Example
5.3.1 Equipment
(1) Adjustable DC power supply model:_____ S/No.:_____
(1) DMM model:_____ S/No.:_____
(1) Non-diffuse light source (pen light)
(1) Yellow LED
(1) Blue LED
(1) IR emitter/detector pair (Lite-On LTE-302 emitter, LTR-301 detector)
(1) 470 Ω resistor ¼ watt actual:_____
(1) 33 k Ω resistor ¼ watt actual:_____
IR Emitter Datasheet: http://optoelectronics.liteon.com/upload/download/DS-50-92-0009/E302.pdf
IR Detector Datasheet: http://optoelectronics.liteon.com/upload/download/DS-50-93-0013/LTR-301.pdf

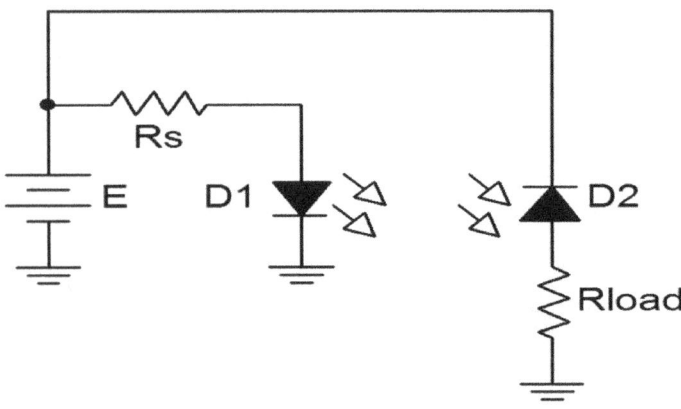

Figure: 5.1 LED Circuit

5.3.2 Procedure
LED as Detector
1. Most LEDs can be used as light detectors. In photovoltaic mode, the output potential is a function of the light level and the make-up of the device (i.e., typically its color). Insert a yellow LED into a protoboard with nothing obstructing it. Place a DMM across it and measure the resulting DC voltage, recording it in Table 5.1 under "Normal".
2. Shade the LED so that minimal light strikes it and measure the resulting voltage. Record the value in Table 5.1 under "Dark".
3. Using the pen light, illuminate the LED from a distance of approximately 10 centimeters, measure and record the voltage in Table 5.1 under "Bright". Also, slowly vary the distance of the pen light from a few centimeters to 20 or so and note what happens to the voltage.
4. Replace the yellow LED with the blue LED and repeat steps 1 through 3.

IR Emitter/Detector Pair
5. Figure 5.1 shows an emitter/detector pair. These devices will emit and detect light at the same wavelength and tend to not produce or detect light at other wavelengths. This aids in avoiding interference. The detector is configured in photoconductive mode. Its current will increase with increasing light level. This current also flows through R_{load} meaning that V_{load} will be proportional to light level.
6. Build the circuit of Figure 5.1 using E = 7 volts, R_s = 470 Ω and R_{load} = 33 kΩ. The emitter diode is denoted with a yellow dot on its case while the detector diode shows a red dot. It is very important that the pair properly be aligned. The bubbles should face each other and cases should be at same height, effectively aiming one bubble at the other. Further, they should only be a few millimeters apart. Finally, the short leads indicate the cathodes.
7. Energize the circuit. Because this pair operates in the infrared, nothing will be apparent to the human eye. Verify that the emitter is operating by measuring the

voltage across it. It should be in the vicinity of 1.1 volts.
8. Measure V_{load} and record the value in Table 5.2.
9. Slip an opaque card such as a thin piece of black plastic or cardboard between the emitter/detector pair. Measure and record V_{load} in Table 5.2.

5.3.3 Result Tables

Table 5.1, Color: RED

Variation	$V_{LED-Yellow}$
Normal	
Dark	
Bright	

Table 5.2 Color: Blue

Variation	V_{Load}
Open	
Blocked	

$V_{LED-Blue}$

5.3.4 Questions

1. What is the effect of light intensity on the LED when used in photovoltaic mode?

   _____
   _____
   _____
   _____

2. What influence does the color of the LED have on the voltage produced when used in photovoltaic mode?

   _____
   _____
   _____
   _____

3. What is the correlation between V_{load} and light level in Figure 5.1? Give at least two examples of where this effect might be put to good use.

   _____
   _____
   _____
   _____

4. Why might an infrared emitter/detector system be used in consumer electronics in place of ordinary visible light emitter/detectors?

   _____
   _____
   _____

CHAPTER SIX
The Zener Diode

6.1 Objective
The objective of this exercise is to examine the operation of the Zener diode and to plot its characteristic curve.

6.2 Theory Overview
When forward biased, the Zener diode behaves similarly to an ordinary switching diode, that is, it incurs a 0.7 volt drop for silicon devices. Unlike a switching diode, the Zener is normally placed in reverse bias. If the circuit potential is high enough, the Zener will exhibit a fixed voltage drop. This is called the Zener potential or V_Z. Manufacturer's specify this voltage with respect to the Zener test current, or I_{ZT}; a point past the knee of the voltage-current curve. That is, if the Zener's current is at least equal to I_{ZT}, then its voltage is approximately equal to the rated V_Z. Above this current, even very large increases in current will produce only very modest changes in voltage. Therefore, for basic circuit analysis, the Zener can be replaced mathematically by a fixed voltage source equal to V_Z. In practice, some series resistance is usually required to limit the current to a value below the Zener's maximum in order to prevent damage.

6.3 Experimental Example
6.3.1 Equipment
(1) Adjustable DC power supply model:_____ S/No.:_____
(1) DMM model:_____ S/No.:_____
(1) Zener diode around 5.1 volts (NZX5V1B, 1N751)
(1) 2.2 k Ω resistor ¼ watt actual: _____
(1) 4.7 k Ω resistor ¼ watt actual: _____
NZX5V1B Datasheet: https://assets.nexperia.com/documents/data-sheet/NZX_SER.pdf

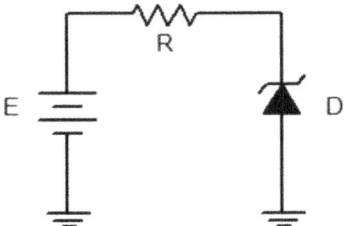

1N751 Datasheet: http://www.digitroncorp.com/Documents/Datasheets/1N746-1N759A,-1N4370-1N4372A.aspx?ext=.pdf

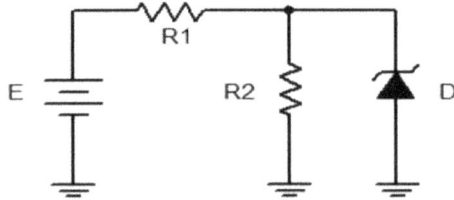

Figure 6.1   Figure 6.2

6.3.2 Procedure
Reverse Curve

1. Consider the circuit of Figure 1 using R = 2.2 kΩ. For any positive value of E the Zener is reverse biased. Until the Zener potential is reached, the diode resistance is effectively infinite and thus no current flows. In this case the voltage across R is zero due to Ohm's law. Consequently, all of E should appear across the Zener. Once the source exceeds the Zener voltage, the remainder of E (i.e. E minus the Zener potential) drops across R. Thus, as E increases, the circulating current increases but the voltage across the zener remains steady.
2. Build the circuit of Figure 6.1 using R = 2.2 kΩ. Set E to 0 volts and measure both the diode's voltage and current and record the results in Table 1. Repeat this process for the remaining source voltages listed.
3. From the data collected in Table 6.1, plot the current versus voltage characteristic of the reverse biased diode. Make sure V_D is the horizontal axis with I_D on the vertical.

Practical Analysis

4. Consider the circuit of Figure 6.2 using R_1 = 2.2 kΩ and R2 = 4.7 kΩ. In general, to analyze circuits like this, first assume that the Zener is out of the circuit and then compute the voltage across R_2 using the voltage divider rule. If the resulting voltage is less than the Zener potential then the Zener is inactive (high resistance) and does not affect the circuit. If, on the other hand, the resulting voltage is greater than the Zener potential then the Zener is active and will limit the voltage across R_2 to V_Z. Via KVL, the remainder of the voltage drops across R_1 and from this the supply current may be determined. This current will then split between R_2 and the Zener. The R_2 current is found using Ohm's law. The Zener current is then found via KCL. Note that for higher and higher values of E, the voltage across (and therefore the current through) R_2 does not change. Instead, all of the "excess" current from the source passes through the Zener.
5. Build the circuit of Figure 6.2 using R_1 = 2.2 kΩ and R_2 = 4.7 kΩ. Set E to 2 volts. Compute the theoretical diode voltage and current, and record them in the first row of Table 6.2. Then measure the diode current and voltage and record in Table 6.2. Finally, compute and record the deviations.
6. Repeat step 5 for the remaining source voltages in Table 6.2.

Computer Simulation

7. Repeat steps 5 and 6 using a simulator, recording the results in Table 6.3.

6.3.3 Result Tables
Table 6.1

E(volts)	V_D	I_D
0		
1		
2		
5		
10		
15		
20		

Table 6.2

E(volts)	$V_{D\ Theory}$	$I_{D\ Theory}$	$V_{D\ Exp.}$	$I_{D\ Exp.}$	% Dev V_D	% Dev I_D
2						
5						
10						
20						

Table 6.3

E(volts)	$V_{D\ Sim}$	$I_{D\ Sim}$
2		
5		
10		
15		
20		

6.3.4 Questions

1. Is it safe to assume that the voltage across a Zener is always equal to the rated V_Z? Why/why not?

   _____
   _____
   _____
   _____
   _____

2. The instantaneous resistance (also known as AC resistance) of a diode may be

approximated by taking the differences between adjacent current-voltage readings. That is, $r_{diode} = \Delta V_{diode}/\Delta I_{diode}$. What is the smallest effective resistance of the Zener using Table 6.1 (show work)?

3. If the circuit of Figure 6.1 had been constructed with the Zener flipped, how would this effect the results recorded in Table 6.1?

4. Assume that a diode with a much higher I_{ZT} rating (say, 100 mA) was used in this exercise. In general, what would the likely outcome be for the circuit of Figure 6.2?

CHAPTER SEVEN A
The Oscilloscope (Tektronix MDO3000)

7.1 Objective
This exercise is of a particularly practical nature, namely, introducing the use of the oscilloscope. The various input scaling, coupling, and triggering settings are examined along with a few specialty features.

7A.2 Theory Overview
The oscilloscope (or simply scope, for short) is arguably the single most useful piece of test equipment in an electronics laboratory. The primary purpose of the oscilloscope is to plot a voltage versus time although it can also be used to plot one voltage versus another voltage, and in some cases, to plot voltage versus frequency. Oscilloscopes are capable of measuring both AC and DC waveforms, and unlike typical
DMMs, can measure AC waveforms of very high frequency (typically 100 MHz or more versus an upper limit of around 1 kHz for a general purpose DMM). It is also worth noting that a DMM will measure the RMS value of an AC sinusoidal voltage, not its peak value.

While the modern digital oscilloscope on the surface appears much like its analog ancestors, the internal circuitry is far more complicated and the instrument affords much greater flexibility in measurement.

Modern digital oscilloscopes typically include measurement aides such as horizontal and vertical cursors or bars, as well as direct readouts of characteristics such as waveform amplitude and frequency. At a minimum, modern oscilloscopes offer two input measurement channels although four and eight channel instruments are increasing in popularity.

Unlike handheld DMMs, most oscilloscopes measure voltages with respect to ground, that is, the inputs are not floating and thus the black, or ground, lead is always connected to the circuit ground or common node. This is an extremely important point as failure to remember this may lead to the inadvertent short circuiting of components during measurement. The standard accepted method of measuring a non-ground referenced potential is to use two probes, one tied to each node of interest, and then setting the oscilloscope to subtract the two channels rather than display each separately. Note that this technique is not required if the oscilloscope has floating inputs (for example, in a handheld oscilloscope). Further, while it is possible to measure non-ground referenced signals by floating the oscilloscope itself through defeating the ground pin on the power cord, this is a safety violation and should not be done.

7A.3 Experimental Example
7A.3.1 Equipment
(1) DC power supply model:_____ S/No.:_____

(1) Function generator model:_____ S/No.:_____
(1) DMM model:_____ S/No.:_____
(1) Oscilloscope, Tektronix MDO 3000 series model:_____ S/No.:_____
Components
(1) 10 kΩ actual:_____
(1) 33 kΩ actual:_____

Schematic Diagrams

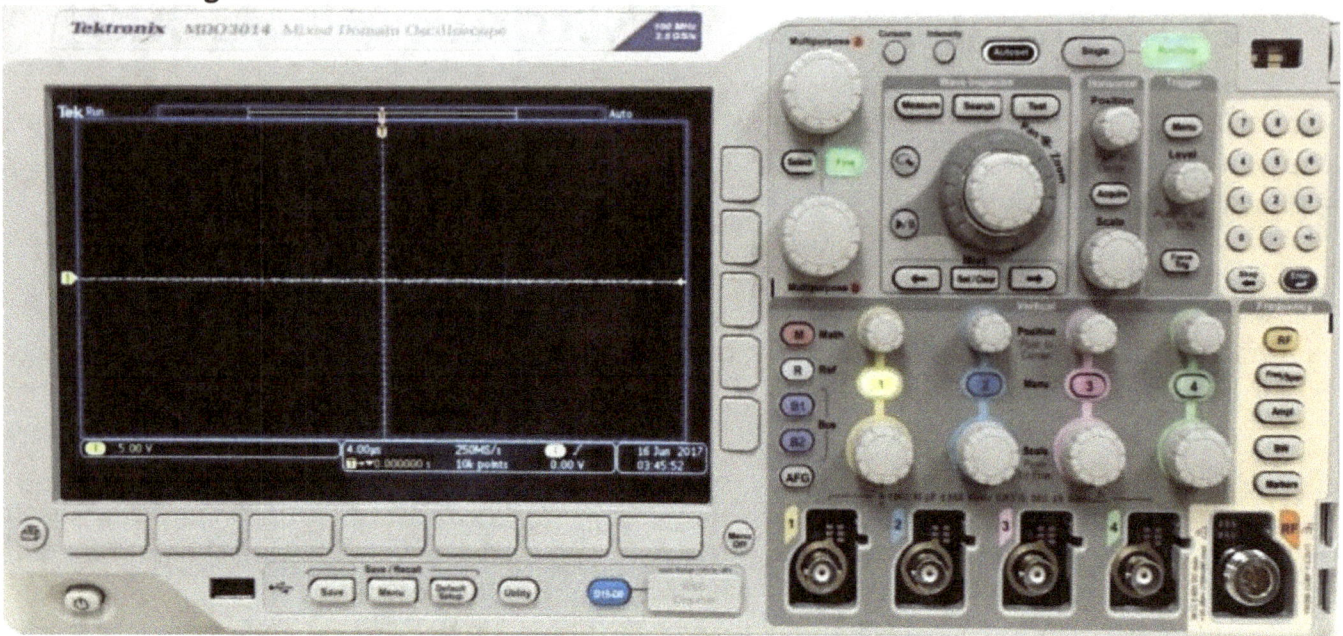

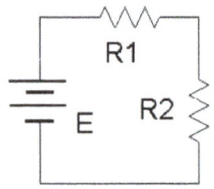

Plate 7A.1

Figure 7A.2

7A.3.2 Procedure
1. Figure 7A.1 is a photo of the face of a Tektronix MDO 3000 series oscilloscope. Compare

this to the bench oscilloscope and identify the following elements:
1) Channel one through four BNC input connectors.
2) RF input connector and settings section.
3) Channel one through four select buttons.
4) Horizontal Scale (i.e., Sensitivity) and Position knobs.
5) Four Vertical Scale (i.e., Sensitivity) and Position knobs.
6) Trigger Level knob.
7) Math and Measure (in Wave Inspector) buttons.
8) Save button (below display).
9) Autoset button.
10) Menu Off button.
2. Note the numerous buttons along the bottom and side of the display screen. These menu buttons are context-sensitive and their function will depend on the most recently selected button or knob. Menus may be removed from the display by pressing the Menu Off button (multiple times for nested menus). Powerup the oscilloscope. Note that the main display is similar to a sheet of graph paper. Each square will have an appropriate scaling factor or weighting, for example, 1 volt per division vertically or 2 milliseconds per division horizontally. Waveform voltages and timings may be determined directly from the display by using these scales.
3. Select the channel one and two buttons (yellow and blue) and also press the Autoset button. (Autoset tries to create reasonable settings based on the input signal and is useful as a sort of "panic button"). There should now be two horizontal lines on the display, one yellow and one blue. These traces may be moved vertically on the display via the associated Position knobs. Also, a trace can be removed by deselecting the corresponding channel button. The Vertical and Horizontal Scale knobs behave in a similar fashion and do not include calibration markings. That is because the settings for these knobs show up on the main display. Adjust the Scale knobs and note how the corresponding values at the bottom of the display change. Voltages are in a 1/2/5 scale sequence while Time is in a 1/2/4 scale sequence.
4. When an input is selected, a menu will pop up allowing control over that input's basic settings. One of the more important fundamental settings on an oscilloscope channel is the Input Coupling. This is controlled via one of the bottom row buttons. There are two choices: AC allows only AC signals through thus blocking DC, and DC allows all signals through (it does not prevent AC).
5. Set the channel one Vertical Scale to 5 volts per division. Set the channel two Scale to 2 volts per division. Set the Time (Horizontal) Scale to 1 millisecond per division. Finally, set the input coupling to DC for both input channels and align the blue and yellow display lines to the center line of the display via the Vertical Position knob (note that pushing the vertical Position knobs will automatically center the trace).
6. Build the circuit of figure 7A. 2 using E=10 V, R1=10 kΩ and R2= 33kΩ. Connect a probe from the channel one input to the power supply (red or tip to the positive terminal, black clip to ground). Connect a second probe from channel two to R2 (again, red or tip to the high side of the resistor and the black clip to ground).
7. The yellow and blue lines should have deflected upward. Channel one should be raised two divisions (2 divisions at 5 volts per division yield the 10 volt source). Using this

method, determine the voltage across R2 (remember, input two should have been set for 2 volts per division). Calculate the expected voltage across R2 using measured resistor values and compare the two in Table 7A.1. Note that it is not possible to achieve extremely high precision using this method (e.g., four or more digits). Indeed, a DMM is often more useful for direct measurement of DC potentials. Double check the results using a DMM and the final column of Table 7A. 1.
8. Select AC coupling for the two inputs. The flat DC lines should drop back to zero. This is because AC Coupling blocks DC. This will be useful for measuring the AC component of a combined AC/DC signal, such as might be seen in an audio amplifier. Set the input coupling for both channels back to DC.
9. Replace the DC power supply with the function generator. Set the function generator for a one volt peak sine wave at 1 kHz and apply it to the resistor network. The display should now show two small sine waves. Adjust the Vertical Scale settings for the two inputs so that the waves take up the majority of the display. If the display is very blurry with the sine waves appearing to jump about side to side, the Trigger Level may need to be adjusted. Also, adjust the Time Scale so that only one or two cycles of the wave may be seen. Using the Scale settings, determine the two voltages (following the method of step 7) as well as the waveform's period and compare them to the values expected via theory, recording the results in Tables 7A. 2 and 3. Also crosscheck the results using a DMM to measure the RMS voltages.
10. To find the voltage across R1, the channel two voltage (voltage across R2) may be subtracted from channel one (E source) via the Math function. Use the red button to select the Math function and create the appropriate expression from the menu (ch1 – ch2). This display shows up in red. To remove a waveform, press its button again. Remove the math waveform before proceeding to the next step.
11. One of the more useful aspects of the oscilloscope is the ability to show the actual wave-shape. This may be used, for example, as a means of determining distortion in an amplifier. Change the wave-shape on the function generator to a square wave, triangle, or other shape and note how the oscilloscope responds. Note that the oscilloscope will also show a DC component, if any, as the AC signal being offset or "riding on the DC". Adjust the function generator to add a DC offset to the signal and note how the oscilloscope display shifts. Return the function generator back to a sine wave and remove any DC offset.
12. It is often useful to take precise differential measurement on a waveform. For this, the bars or cursors are useful. Select the Cursors button toward the top of the oscilloscope. From the menu on the display, select Vertical Bars. Two vertical bars will appear on the display (it is possible that one or both could be positioned off the main display). They may be moved left and right via the multipurpose knobs (next to the Cursors button). The Select button toggles between independent and tandem cursor movement. Readout of the bar values will appear in the upper portion of the display. They indicate the positions of the cursors, i.e., the location where they cross the waveform. Vertical Bars are very useful for obtaining time information as well as amplitudes at specific points along the wave. A similar function is the Horizontal Bars which are particularly useful for determining amplitudes. Try the Horizontal Bars by selecting them via the Cursors menu again (holding the Cursors button will bring up the menu).

13. For some waveform parameters, automatic readings are available. These are accessed via the Measure button. Press Measure, select Add Measurement, and page through the various options using the multipurpose b knob. Select Frequency. Note that a small readout of the frequency will now appear on the display. Multiple measurements are possible simultaneously. Important: There are specific limits on the proper usage of these measurements. If the guidelines are not followed, erroneous values may result. Always perform an approximation via the Scale factor and divisions method even when using an automatic measurement!
14. Finally, a snap-shot of the screen may be saved for future work using the USB port and a USB memory stick via the Save Menu button. The popup menu has options for saving the image as well as the trace data or setup info. Select Save Screen Image to save a bit mapped graphics file that can be used as is or processed further in a graphics program (for example, inverting the colors for printing). The .PNG format is recommended.

7A.3.3 Result Tables

Table 7A.1

V_{R2}	Scale (V/Div)	Number of Divisions	Voltage Scope	Voltage DMM
Oscilloscope				
Theory	X	X		

Table 7A.2

	Scale (V/Div)	Number of Divisions	Voltage Peak	Voltage RMS
E Oscilloscope				
E Theory	X	X		
V_{R2} Oscilloscope				
V_{R2} Theory	X	X		

Table 7A.3

	Scale (S/Div)	Number of Divisions	Period	Frequency
E Oscilloscope				
E Theory	X	X		

CHAPTER SEVEN B
The Oscilloscope (Tektronix TDS 3000)

7B.1 Objective
This exercise is of a particularly practical nature, namely, introducing the use of the oscilloscope. The various input scaling, coupling, and triggering settings are examined along with a few specialty features.

7B.2 Theory Overview
The oscilloscope (or simply scope, for short) is arguably the single most useful piece of test equipment in an electronics laboratory. The primary purpose of the oscilloscope is to plot a voltage versus time although it can also be used to plot one voltage versus another voltage, and in some cases, to plot voltage versus frequency. Oscilloscopes are capable of measuring both AC and DC waveforms, and unlike typical DMMs, can measure AC waveforms of very high frequency (typically 100 MHz or more versus an upper limit of around 1 kHz for a general purpose DMM). It is also worth noting that a DMM will measure the RMS value of an AC sinusoidal voltage, not its peak value.

While the modern digital oscilloscope on the surface appears much like its analog ancestors, the internal circuitry is far more complicated and the instrument affords much greater flexibility in measurement.

Modern digital oscilloscopes typically include measurement aides such as horizontal and vertical cursors or bars, as well as direct readouts of characteristics such as waveform amplitude and frequency. At a minimum, modern oscilloscopes offer two input measurement channels although four and eight channel instruments are increasing in popularity.

Unlike handheld DMMs, most oscilloscopes measure voltages with respect to ground, that is, the inputs are not floating and thus the black, or ground, lead is always connected to the circuit ground or common node. This is an extremely important point as failure to remember this may lead to the inadvertent short circuiting of components during measurement. The standard accepted method of measuring a non-ground referenced potential is to use two probes, one tied to each node of interest, and then setting the oscilloscope to subtract the two channels rather than display each separately. Note that this technique is not required if the oscilloscope has floating inputs (for example, in a handheld oscilloscope). Further, while it is possible to measure non-ground referenced signals by floating the oscilloscope itself through defeating the ground pin on the power cord, this is a safety violation and should not be done.

7B.3 Experimental Example
7B.3.1 Equipment
(1) DC power supply model:_____ S/No.:_____
(1) Function generator model:_____ S/No.:_____

(1) DMM model:_____ S/No.:_____
(1) Oscilloscope, Tektronix TDS 3000 series model:_____ S/No.:_____
Components
(1) 10 kΩ actual:_____
(1) 33 kΩ actual:_____

Schematic Diagrams

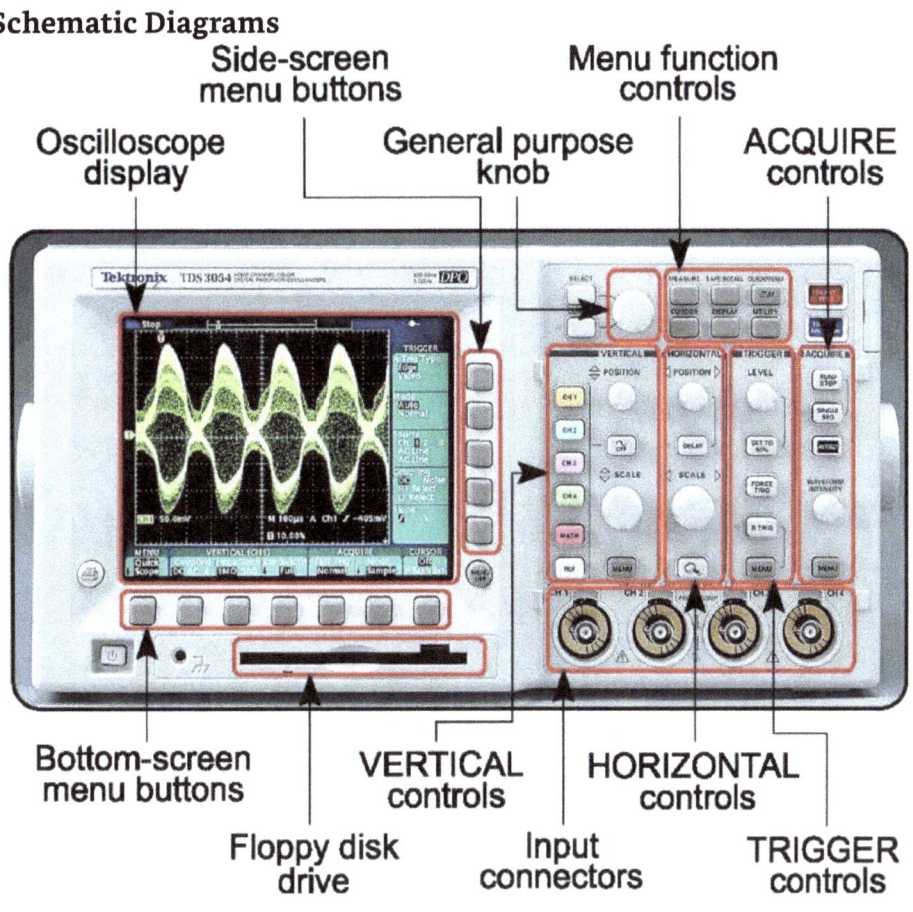

Figure 7B.1a: (Four channel version shown)

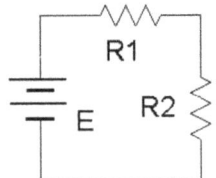

Figure 7B.2

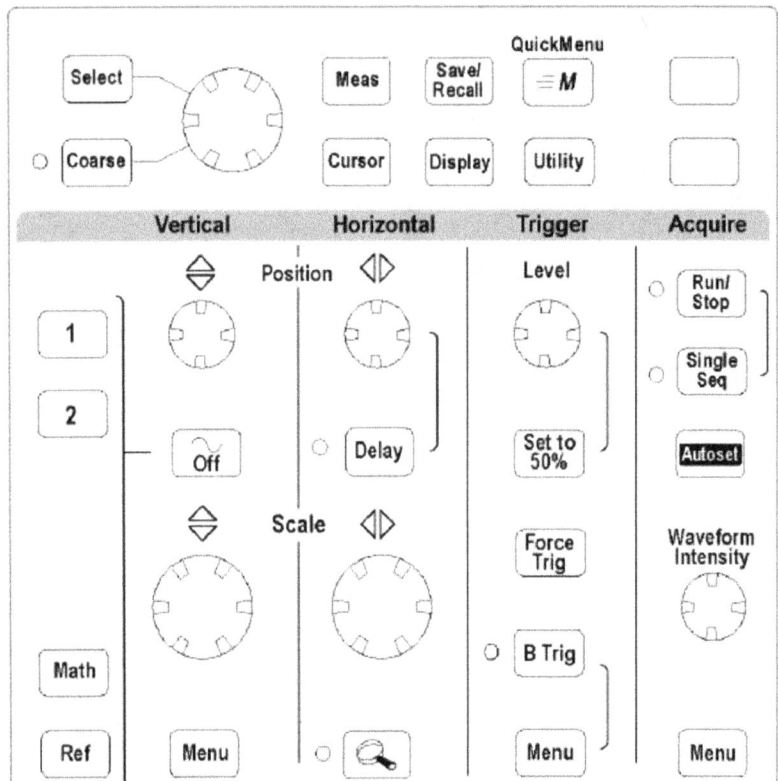

Figure 7B.1b

7B.3.2 Procedure

1. Figure 7B.1 is an outline of the main face of a Tektronix TDS 3000 series oscilloscope. Compare this to the bench oscilloscope and identify the following elements:

- ✓ Channel one and two BNC input connectors.
- ✓ Trigger BNC input connector.
- ✓ Channel one and two select buttons.
- ✓ Horizontal sensitivity (or Scale) and Position knobs.
- ✓ Vertical sensitivity (or Scale) and Position knobs.
- ✓ Trigger Level knob.
- ✓ Quick Menu button.
- ✓ Print/Save button.
- ✓ Autoset button.

2. Note the numerous buttons along the bottom and side of the display screen. These buttons are context-sensitive and their function will depend on the mode of operation of the oscilloscope. Power up the oscilloscope and select the Quick Menu button. Notice that the functions are listed next to the buttons. This is a very useful menu and serves as a good starting point for most experiment setups.

Note that the main display is similar to a sheet of graph paper. Each square will have an appropriate scaling factor or weighting, for example, 1 volt per division vertically or 2 milliseconds per division horizontally. Waveform voltages and timings may be determined directly from the display by using these scales.

3. Select the channel one and two buttons (yellow and blue) and also select the Autoset button. (Autoset tries to create reasonable settings based on the input signal and is useful as a sort of "panic button").

There should now be two horizontal lines on the display, one yellow and one blue. They may be moved via the Position knob. The Position knob moves the currently selected input (select the channel buttons alternately to toggle back and forth between the two inputs). The Vertical and Horizontal Scale knobs behave in a similar fashion and do not include calibration markings. That is because the settings for these knobs show up on the main display. Adjust the Scale knobs and note how the corresponding values in the display change. Voltages are in a 1/2/5 scale sequence while
Time is in a 1/2/4 scale sequence.

4. One of the more important fundamental settings on an oscilloscope is the Input Coupling.

This is controlled via one of the bottom row buttons. There are three choices: Ground removes the input thus showing a zero reference, AC allows only AC signals through thus blocking DC, and DC allows all signals through (it does not prevent AC).

5. Set the channel one Vertical Scale to 5 volts per division. Set the channel two Scale to 2 volts per division. Set the Time (Horizontal) Scale to 1 millisecond per division. Finally, set the input coupling to Ground for both input channels and align the blue and yellow display lines to the center line of the display via the Vertical Position knob.

6. Build the circuit of Figure 7B.2 using E=10 V, R1=10 kW and R2= 33kW. Connect a probe from the channel one input to the power supply (red or tip to plus, black clip to ground). Connect a second probe from channel two to R2 (again, red or tip to the high side of the resistor and the black clip to ground).

7. Switch both inputs to DC coupling. The yellow and blue lines should have deflected upward. Channel one should be raised two divisions (2 divisions at 5 volts per division yield the 10 volt source). Using this method, determine the voltage across R2 (remember, input two should have been set for 2 volts per division). Calculate the expected voltage across R2 using measured resistor values and compare the two in Table 7B.1. Note that it is not possible to achieve extremely high precision using this method (e.g., four or more digits). Indeed, a DMM is often more useful for direct measurement of DC potentials. Double check the results using a DMM and the final column of Table 7B.1.

8. Select AC coupling for the two inputs. The flat DC lines should drop back to zero. This is because AC coupling blocks DC. This will be useful for measuring the AC component of a combined AC/DC signal, such as might be seen in an audio amplifier. Set the input coupling for both channels back to DC.

9. Replace the DC power supply with the function generator. Set the function generator for a one volt peak sine wave at 1 kHz and apply it to the resistor network. The display should now show two small sine waves. Adjust the Vertical Scale settings for the two inputs so that the waves take up the majority of the display. If the display is very blurry with the sine waves appearing to jump about side to side, the Trigger Level may need to be adjusted. Also, adjust the Time Scale so that only one or two cycles of the wave may be seen. Using the Scale settings, determine the two voltages (following the method of step 7) as well as the waveform's period and compare them to the values expected via theory, recording the results in Tables 7B.2 and 3. Also crosscheck the results using a DMM to measure the RMS voltages.

10. To find the voltage across R1, the channel two voltage (voltage across R2) may be subtracted from channel one (E source) via the Math function. Use the red button to select the Math function and create the appropriate expression from the menu (ch1 − ch2). This display shows up in red. To remove a waveform, select it and then select Off. Remove the math waveform before proceeding to the next step.

11. One of the more useful aspects of the oscilloscope is the ability to show the actual wave-shape. This may be used, for example, as a means of determining distortion in an amplifier. Change the wave-shape on the function generator to a square wave, triangle, or other shape and note how the oscilloscope responds. Note that the oscilloscope will also show a DC component, if any, as the AC signal being offset or "riding on the DC". Adjust the function generator to add a DC offset to the signal and note how the oscilloscope display shifts. Return the function generator back to a sine wave and remove any DC offset.

12. It is often useful to take precise differential measurement on a waveform. For this, the bars or cursors are useful. Select the Cursor button toward the top of the oscilloscope. From the menu on the display, select Vertical Bars. Two vertical bars will appear on the display (it is possible that one or both could be positioned off the main display). They may be moved left and right via the function knob (next to the Cursor button). The Select button toggles between the two cursors. A read-out of the bar values will appear in the upper portion of the display. They indicate the positions of the cursors, i.e. the location where they cross the waveform. Vertical Bars are very useful for obtaining time information as well as amplitudes at specific points along the wave. A similar function is the Horizontal Bars which are particularly useful for determining amplitudes. Try the Horizontal Bars by selecting them via the Cursor button again.
13. For some waveform parameters, automatic readings are available. These are accessed via the **Meas** (Measurement) button. Select **Meas** and page through the various options. Select *Frequency*. Note that a small readout of the frequency will now appear on the display. Up to four measurements are possible simultaneously. **Important**: There are specific limits on the proper usage of these measurements. If the guidelines are not followed, erroneous values may result. **Always** perform an approximation via the Scale factor and divisions method even when using an automatic measurement.
14. Finally, a snap-shot of the screen may be saved for future work using the floppy disk drive via the Printer button. The result will be a bit mapped graphics file that can be used as is or processed further in a graphics program (for example, inverting the colors for printing).

7B.3.3 Result Tables
Table 7B.1

V_{R2}	Scale (V/Div)	Number of Divisions	Voltage Scope	Voltage DMM
Oscilloscope				
Theory	X	X		

Table 7B.2

	Scale (V/Div)	Number of Divisions	Voltage Peak	Voltage RMS
E Oscilloscope				
E Theory	X	X		
V_{R2} Oscilloscope				
V_{R2} Theory	X	X		

Table 7B.3

	Scale (S/Div)	Number of Divisions	Period	Frequency

E Oscilloscope				
E Theory	X	X		

CHAPTER SEVEN C
The Oscilloscope (GWInstek 2000)

7C.1 Objective
This exercise is of a particularly practical nature, namely, introducing the use of the oscilloscope. The various input scaling, coupling, and triggering settings are examined along with a few specialty features.

7C.2 Theory Overview
The oscilloscope (or simply scope, for short) is arguably the single most useful piece of test equipment in an electronics laboratory. The primary purpose of the oscilloscope is to plot a voltage versus time although it can also be used to plot one voltage versus another voltage, and in some cases, to plot voltage versus frequency. Oscilloscopes are capable of measuring both AC and DC waveforms, and unlike typical DMMs, can measure AC waveforms of very high frequency (typically 100 MHz or more versus an upper limit of around 1 kHz for a general purpose DMM). It is also worth noting that a DMM will measure the RMS value of an AC sinusoidal voltage, not its peak value.

While the modern digital oscilloscope on the surface appears much like its analog ancestors, the internal circuitry is far more complicated and the instrument affords much greater flexibility in measurement. Modern digital oscilloscopes typically include measurement aides such as horizontal and vertical cursors or bars, as well as direct readouts of characteristics such as waveform amplitude and frequency. At a minimum, modern oscilloscopes offer two input measurement channels although four and eight channel instruments are increasing in popularity. Unlike handheld DMMs, most oscilloscopes measure voltages with respect to ground, that is, the inputs are not floating and thus the black, or ground, lead is always connected to the circuit ground or common node. This is an extremely important point as failure to remember this may lead to the inadvertent short circuiting of components during measurement. The standard accepted method of measuring a non-ground referenced potential is to use two probes, one tied to each node of interest, and then setting the oscilloscope to subtract the two channels rather than display each separately. Note that this technique is not required if the oscilloscope has floating inputs (for example, in a handheld oscilloscope). Further, while it is possible to measure non-ground referenced signals by floating the oscilloscope itself through defeating the ground pin on the power cord, this is a safety violation and should not be done.

7C.3 Experimental Example
7C.3.1 Equipment
(1) DC power supply model:_____ S/No.:_____
(1) Function generator model:_____ S/No.:_____
(1) DMM model:_____ S/No.:_____
(1) Oscilloscope, GWInstek 2000 series model:_____ S/No.:_____

Components
(1) 10 kΩ actual:_____
(1) 33 kΩ actual:_____

Schematic Diagrams

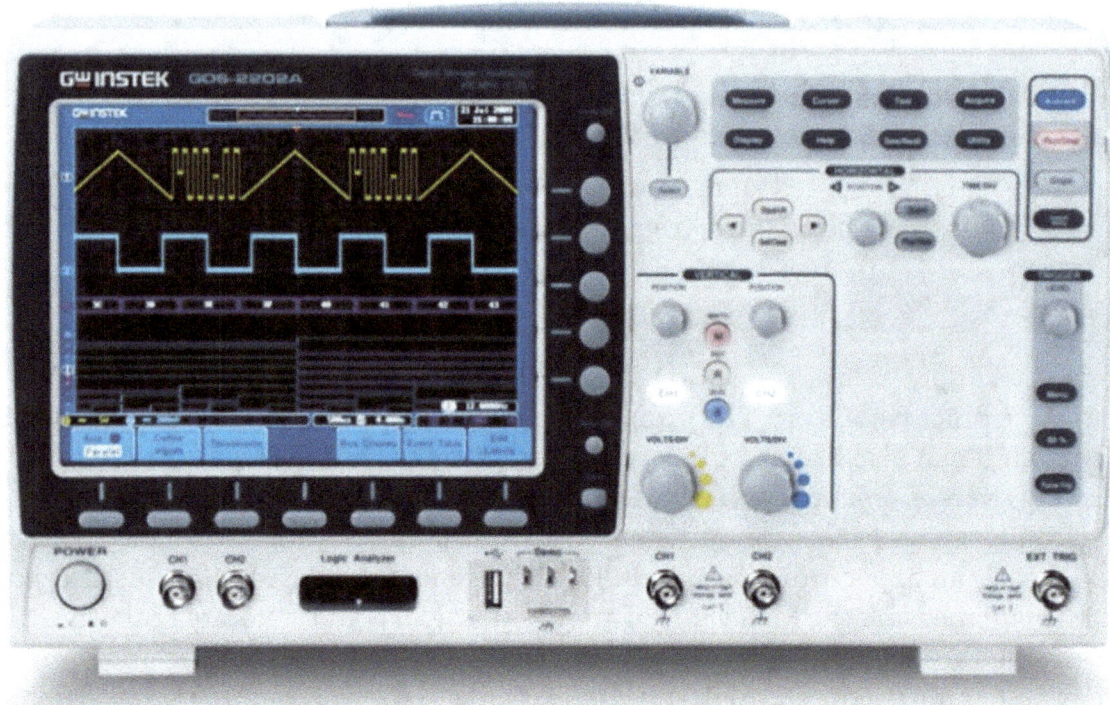

Figure 7C.1a

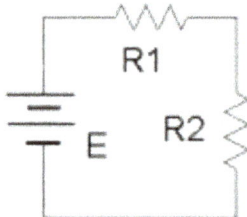

Figure 7C.2

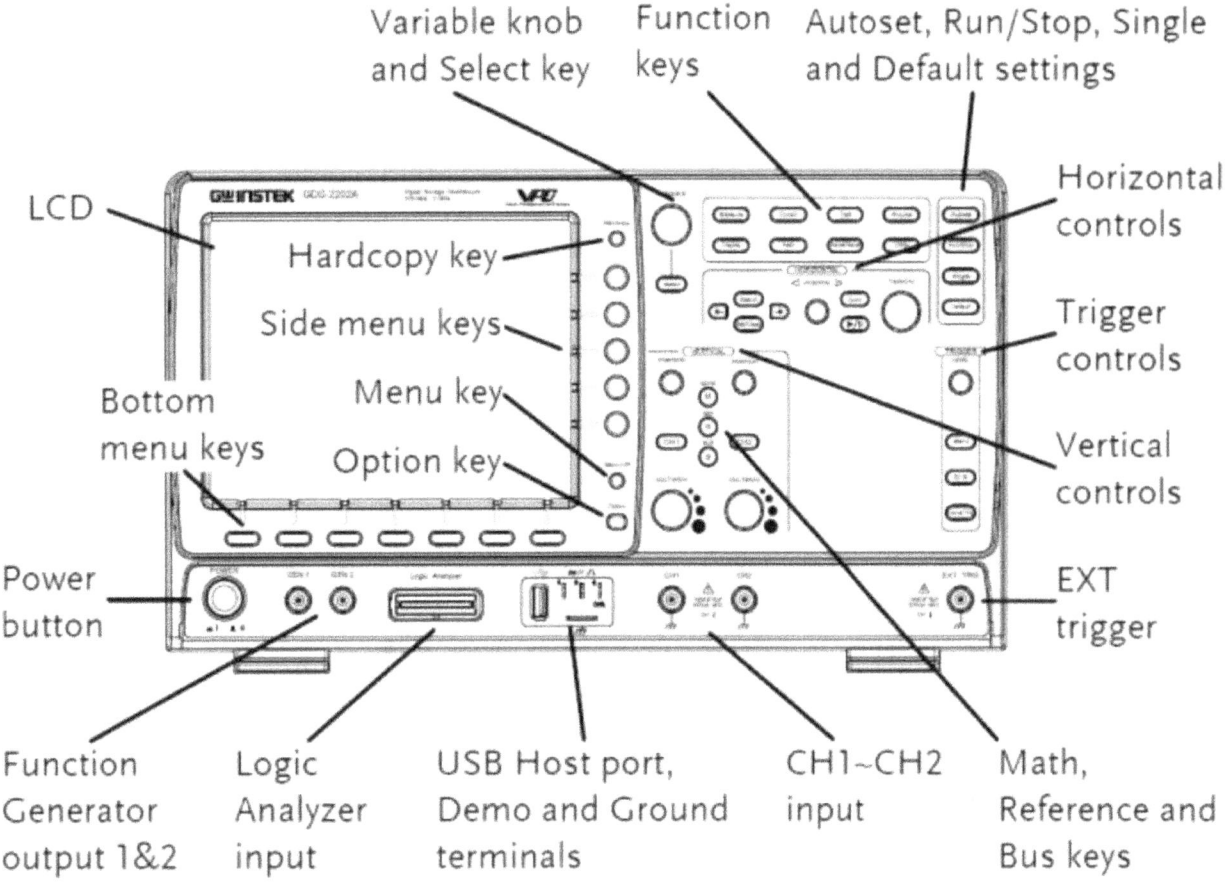

Figure 7C.1b

7C.3.2 Procedure

1. Figure 1 is an outline of the main face of a GWInstek 2000 series oscilloscope. Compare this to the bench oscilloscope and identify the following elements:
   - ✓ Channel one and two BNC input connectors.
   - ✓ Trigger BNC input connector.
   - ✓ Horizontal sensitivity (Time/Div) and Position knobs.
   - ✓ Channel one and two Select buttons (lighted style).
   - ✓ Channel one and two Vertical sensitivity (Volts/Div) and Position knobs.
   - ✓ Trigger Level knob.
   - ✓ Function keys, including Cursor and Measure.
   - ✓ Hardcopy button.
   - ✓ Autoset button.
   - ✓ Math button.
   - ✓ USB port.
2. Note the numerous buttons along the bottom and side of the display screen. These buttons are context-sensitive and their function will depend on the mode of operation of the oscilloscope. Power up the oscilloscope. Notice that the functions are listed next to the buttons. Note that the main display is similar to a sheet of graph paper. Each square will have an appropriate scaling factor or weighting, for example, 1 volt per division vertically or 2 milliseconds per division horizontally. Waveform voltages and timings may be determined directly from the display by using these scales.

3. Depress the channel one and two Select buttons (they should light) and also select the Autoset button. (Autoset tries to create reasonable settings based on the input signal and is useful as a sort of "panic button"). There should now be two horizontal lines on the display, one yellow and one blue. They may be moved via the Position knob. The Position knobs move the associated input. The Vertical and Horizontal Scale knobs behave in a similar fashion and do not include calibration markings. That is because the settings for these knobs show up on the main display. Adjust the Scale knobs and note how the corresponding values in the display change. Voltages and Time base use a 1/2/5 scale sequence.

4. One of the more important fundamental settings on an oscilloscope is the Input Coupling. This is controlled via one of the bottom row buttons. There are three choices: Ground removes the input thus showing a zero reference, AC allows only AC signals through thus blocking DC, and DC allows all signals through (it does not prevent AC).

5. Set the channel one Vertical Scale to 5 volts per division. Set the channel two Scale to 2 volts per division. Set the Time (Horizontal) Scale to 1 millisecond per division. Finally, set the input coupling to Ground for both input channels and align the blue and yellow display lines to the center line of the display via the Vertical Position knobs.

6. Build the circuit of Figure 7C.2 using E=10 V, R1=10 kΩ and R2= 33kΩ. Connect a probe from the channel one input to the power supply (red or tip to plus, black clip to ground). Connect a second probe from channel two to R2 (again, red or tip to the high side of the resistor and the black clip to ground).

7. Switch both inputs to DC coupling. The yellow and blue lines should have deflected upward. Channel one should be raised two divisions (2 divisions' times 5 volts per division yields the 10 volt source). Using this method, determine the voltage across R2 (remember, input two should have been set for 2 volts per division). Calculate the expected voltage across R2 using measured resistor values and compare the two in Table 7C.1. Note that it is not possible to achieve extremely high precision using this method (e.g., four or more digits). Indeed, a DMM is often more useful for direct measurement of DC potentials. Double check the results using a DMM and the final column of Table 7C.1.

8. Select AC coupling for the two inputs. The flat DC lines should drop back to zero. This is because AC coupling blocks DC. This will be useful for measuring the AC component of a combined AC/DC signal, such as might be seen in an audio amplifier. Set the input coupling for both channels back to DC.

9. Replace the DC power supply with the function generator. Set the function generator for a one volt peak sine wave at 1 kHz and apply it to the resistor network. The display should now show two small sine waves. Adjust the Vertical Scale settings for the two inputs so that the waves take up the majority of the display. If the display is very blurry with the sine waves appearing to jump about side to side, the Trigger Level may need to be adjusted. Also, adjust the Time Scale so that only one or two cycles of the wave may be seen. Using the Scale settings, determine the two voltages (following the method of step 7) as well as the waveform's period and compare them to the values expected via theory, recording the results in Tables 7C.2 and 3. Also crosscheck the results using a DMM to measure the RMS voltages.

10. To find the voltage across R1, the channel two voltages (voltage across R2) may be subtracted from channel one (E source) via the Math function. Use the red button to select

the Math function and create the appropriate expression from the menu (ch1 − ch2). This display shows up in red. To remove a waveform, simply deselect it (depress the associated button). Remove the math waveform before proceeding to the next step.

11. One of the more useful aspects of the oscilloscope is the ability to show the actual wave-shape. This may be used, for example, as a means of determining distortion in an amplifier. Change the wave-shape on the function generator to a square wave, triangle, or other shape and note how the oscilloscope responds. Note that the oscilloscope will also show a DC component, if any, as the AC signal being offset or "riding on the DC". Adjust the function generator to add a DC offset to the signal and note how the oscilloscope display shifts. Return the function generator back to a sine wave and remove any DC offset.

12. It is often useful to take precise differential measurement on a waveform. For this, the bars or cursors are useful. Select the Cursor button toward the top of the oscilloscope. From the menu on the display, select Vertical. Two vertical bars will appear on the display (it is possible that one or both could be positioned off the main display). They may be moved left and right via the Variable knob (next to the Cursor button). The Select button toggles between the two cursors. A read out of the bar values will appear in the upper portion of the display. They indicate the positions of the cursors, i.e. the location where they cross the waveform. Vertical Bars are very useful for obtaining time information as well as amplitudes at specific points along the wave. A similar function is the Horizontal Bars which are particularly useful for determining amplitudes. Try the Horizontal Bars by selecting them via the Cursor button again.

13. For some waveforms parameters, automatic readings are available. These are accessed via the Measure button. Select Measure and page through the various options. Select *Frequency*. Note that a small readout of the frequency will now appear on the display. Now try RMS and compare the result to that given by the DMM earlier. Note that several measurements are possible simultaneously. **Important**: There are specific limits on the proper usage of these measurements. If the guidelines are not followed, erroneous values may result. **Always** perform an approximation via the Scale factor and divisions method even when using an automatic measurement.

14. Finally, a snap-shot of the screen may be saved for future work using the USB port and a USB memory stick via the Hardcopy button. The result will be a bit mapped graphics file that can be used as is (see below) or processed further in a graphics program (for example, inverting the colors for printing).

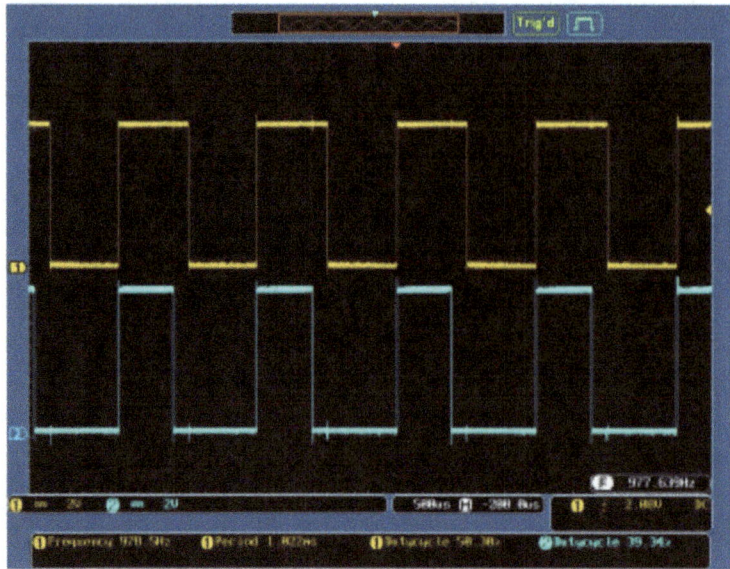

Figure 7C.3

7C.3.3 Result Tables

Table 7B.1

V_{R2}	Scale (V/Div)	Number of Divisions	Voltage Scope	Voltage DMM
Oscilloscope				
Theory	X	X		

Table 7B.2

	Scale (V/Div)	Number of Divisions	Voltage Peak	Voltage RMS
E Oscilloscope				
E Theory	X	X		
V_{R2} Oscilloscope				
V_{R2} Theory	X	X		

Table 7B.3

	Scale (S/Div)	Number of Divisions	Period	Frequency
E Oscilloscope				
E Theory	X	X		

CHAPTER EIGHT
Diode Clippers and Clampers

8.1 Objective
The performance and operation of diode based clipper and clamper circuits are examined in this exercise.
Items of interest include the programmability of the clipper and non-ideal effects in both circuits.

8.2 Theory Overview
The function of the clipper is to ensure that the input waveform never exceeds a certain peak value. This may be a protective function, that is, a large signal might damage a following circuit, but it may be used for other reasons, for example, ensuring that signal overage is never reached. Clipping circuits do not have to be symmetrical. In other words, the positive and negative limits do not have to have the same magnitude. While clippers can be designed using Zener diodes, the biased clipper offers the advantage of infinite variability of the limit point. In contrast, Zener based clippers are limited by the available standard zener potentials and cannot be set to new values without replacement of the Zeners.

Clamper circuits are designed to provide a DC level shift. Typically this means shifting a waveform vertically so that the entire waveform is either positive or negative with one peak now residing at zero.

8.3 Experimental Example
8.3.1 Equipment
 (1) Dual channel oscilloscope model:_____ S/No.:_____
(1) Function generator model:_____ S/No.:_____
(1) Dual adjustable DC power supply model:_____ S/No.:_____
(2) Signal diodes (1N4148, 1N914)
(1) 10 k Ω resistor ¼ watt actual: _____
(1) 100 k Ω resistor ¼ watt actual: _____
(1) 100 nF capacitor 25 volt actual: _____
1N4148/1N914 Datasheet: https://www.onsemi.com/pub/Collateral/1N914A-D.pdf

Schematic Diagrams

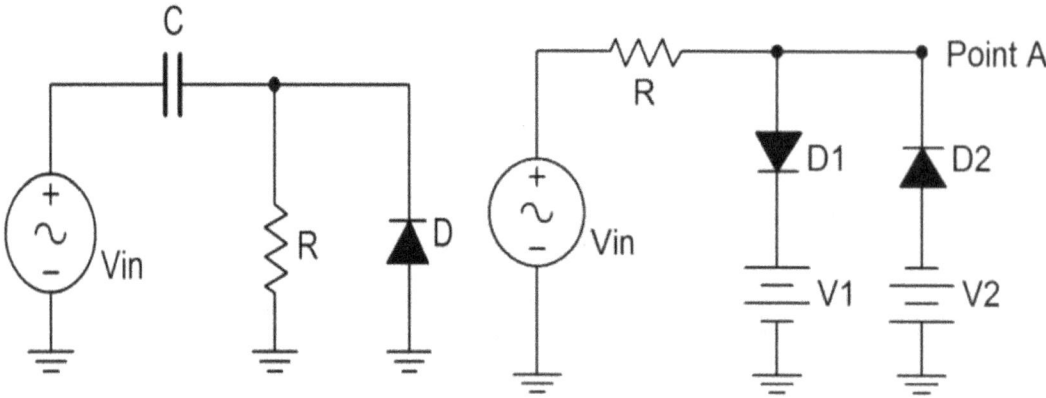

Figure 8.1   Figure 8.2

8.3.2 Procedure
Clipper
1. Consider the circuit of Figure 8.1 using Vin = 5 volts peak at 1 kHz, R = 10 kΩ, and both V1 and V2 set to 0 volts. For any positive signal over approximately 0.7 volts, D1 will turn on and limit the output voltage at Point A to 0.7 volts. A similar situation occurs with D2 and negative signals. A sine wave input over several volts will clipped at ± 0.7 volts resulting in a squared off wave.
2. Build the circuit of Figure 1 using R = 10 kΩ, and both V1 and V2 set to 0 volts. Set Vin to a 5 volt peak sine at 1 kHz. Place one oscilloscope probe at the input and the second at Point A. Make sure the scope inputs are DC coupled. Record the positive and negative peak values of the resulting output waveform in Table 8.1. Also save an image of the scope showing both the input and output waveforms.
3. Set V1 to 2 V_{DC} and V2 to 3 V_{DC}, and repeat step 2

Clamper
4. Consider the circuit of Figure 8.2 using Vin = 5 volts peak at 1 kHz, R = 100 kΩ and C = 100 nF. Determine the input signal's period and the RC time constant. Record these values in Table 2. For a large time constant, the capacitor voltage can be thought of as stable, basically a DC voltage. Initially, this voltage is zero. On the positive input half-wave, the diode is reverse biased and all of the signal drops across the resistor. On negative portions though, the diode is forward biased, limiting the output voltage to within one diode drop of ground. Also, the capacitor will begin to charge, eventually reaching the peak voltage of the input. This potential will add a DC offset to the input signal resulting in a clamped output.
5. Build the circuit of Figure 8.2 using R = 100 kΩ and C = 100 nF. Set Vin to a 5 volt peak sine at 1 kHz. Set the scope inputs to DC coupled and apply the probes to the input and output points. Record the positive and negative peak values of the output waveform in Table 8.3.
6. Reverse the diode and repeat step 5.

Computer Simulation
7. Repeat the clamper procedure of steps 5 and 6 using a simulator, recording the results in Table 8.4.

8.3.3 Result Tables

Table 8.1

Variation	V_{out} Positive Peak
No Bias	
With Bias	

Table 8.2

Input Period	
RC Time Constant	

V_{out} Negative Peak	

Table 8.3

Diode Polarity	V_{out} Positive Peak	V_{out} Negative Peak
Original		
Reverse		

Table 8.4

Diode Polarity	$V_{out-sim}$ Positive Peak	$V_{out-sim}$ Negative Peak
Original		
Reverse		

8.3.4 Questions

1. Are the clipping thresholds of the circuit of Figure equal to the bias voltages? Why/why not?

   _____
   _____
   _____
   _____
   _____

2. If the magnitudes of V1 and V2 in Figure 1 had been reversed, what would the output waveform look like? Would the peaks of the output have changed?

   _____
   _____
   _____
   _____

3. What would the output waveform look like if D2 and V2 had been omitted in Figure 8.1?

   _____
   _____

4. How accurate is the clamping effect of the circuit of Figure 2? Are there any non-ideal effects? What is the effect of reversing the diode polarity in Figure 8.2?

5. What would the output waveform look like if a much smaller capacitor had been used?

CHAPTER NINE
Half-wave Rectifier

9.1 Objective
The goal of this exercise is to investigate the ideal versus real operation of a basic half-wave rectifier. The effects of a filtering capacitor are included.

9.2 Theory Overview
The primary function of a rectifier is to turn an incoming AC waveform into a pulsating DC waveform.

This can be achieved by simply blocking one of the two polarities from reaching the load. This is called a half-wave rectifier. A rectifier might be used as part of an AC to DC power supply but might also be used as part of a signal processing system. Ideally, a diode will behave as either an open or a closed switch depending on the polarity of the applied signal. This means that one polarity can be allowed through while the other can be blocked, perfect for this situation. In reality, the diode will require a forward turn on potential which results in a portion of the allowed signal being lost (e.g., the first 0.7 volts will be lost when using a silicon device). This fact makes a simple diode rectifier ineffective when used with very small signal amplitudes. Finally, in order to "fill the gap" where the blocked polarity would have been, a capacitor can be used to store some of the energy at the peak to be released during the gap. The higher the capacitance value, the more effective the smoothing will be.

9.3 Experimental Example
9.3.1 Equipment
(1) Dual channel oscilloscope model:_____ S/No.:_____
(1) Function generator model:_____ S/No.:_____
(1) DMM model:_____ S/No.:_____
(1) Signal diode (1N914, 1N4148)
(1) 10 k Ω resistor ¼ watt actual: _____
(1) 22 nF capacitor actual: _____
(1) 470 nF capacitor actual: _____
1N4148/1N914 Datasheet: https://www.onsemi.com/pub/Collateral/1N914A-D.pdf

Schematic Diagrams

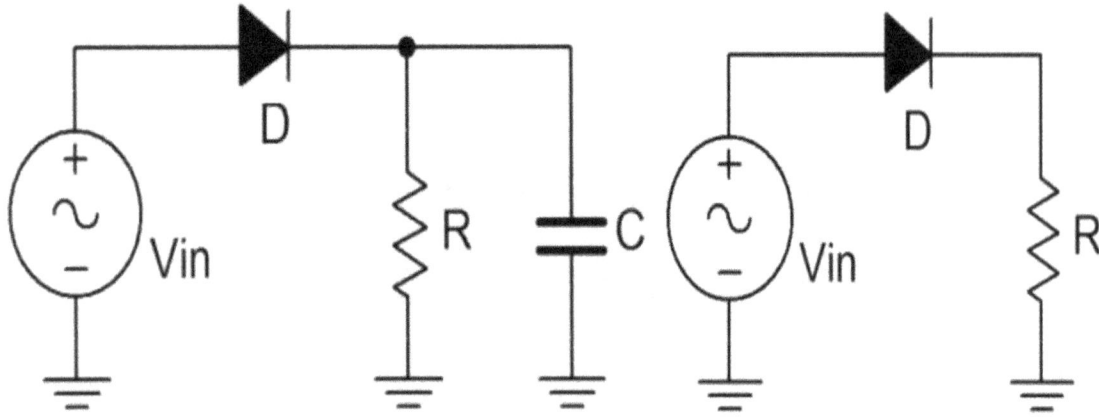

Figure 9.1   Figure 9.2

9.3.2 Procedure
Simple Rectifier
1. Consider the circuit of Figure 9.1. For an input voltage significantly larger than 0.7 volts, the diode will be forward biased for the positive half of the input sine wave. Therefore, all of the input signals (Less 0.7 volts) will appear across the load resistor, R. Conversely, during the negative polarity of the input, the diode will be open, thus blocking any current and producing no voltage across the load.
2. Build the circuit of Figure 1 using Vin = 10 volts peak at 1 kHz and R = 10 kΩ. Set the oscilloscope inputs to DC coupled. Place one oscilloscope probe across the input generator and a second probe across the load resistor. Record the peak amplitude of the output load waveform in Table 9.1. Also, save an image of the scope trace showing both the input and output waveforms.
3. Measure the load voltage with the DMM (DC volts) and record this in Table 9.1.
4. Reverse the diode and repeat steps 2 and 3.
5. Reverse the diode so that it is back to the original orientation. Reduce the input to 800 mV peak and repeat step 2.

Filter Capacitor
6. The circuit of Figure 9.2 adds a filtering capacitor across the load. This should help to "fill the gaps" created by the missing portions of the waveform. Build the circuit of Figure 9.2 using Vin = 10 volts peak at 1 kHz, R = 10 kΩ and C = 22 nF. Making sure that the scope inputs are DC coupled, place scope probes across the input and load, and capture the resulting image. Record the peak value in Table 9.2.
7. Measure the load voltage with the DMM (DC volts) and record this in Table 9.2.
8. Replace C with the 470 nF capacitor and repeat steps 5 and 6.

Computer Simulation
9. Perform a Transient Analysis simulation of the circuits shown in Figures 9.1 and 2, and compare the resulting waveforms to those captured from the oscilloscope.

9.3.3 Result Tables
Table 9.1

Variation	V_{load} peak	DMM DCV

Original		
Reversed Diode		
800 mV Input		X

Table 9.2

Variation	V_{load} peak	DMM DCV
Capacitor		
22 nF		
470 nF		

9.3.4 Questions

1. How well do the peak values of the load voltage track the peak values of the input voltage?

2. What are the limits of the half-wave circuit at rectifying small amplitude signals? What might be done to improve its effectiveness?

3. What is the effect of reversing the orientation of the diode? How does this affect the DC value measured by the DMM?

4. What is the effect of adding capacitance to the circuit? How does this affect the DC value measured by the DMM?

5. How would the waveforms differ if the oscilloscope inputs had been AC coupled

instead of DC coupled?

CHAPTER TEN
The Transformer

10.1 Objective
The objective of this exercise is to introduce the power transformer. Turns ratio and its effects on primary secondary voltage and current are of prime importance. The effect of loading will also be examined.

10.2 Theory Overview
A power transformer is used to change an AC voltage from one amplitude to another, ideally without power loss. This is accomplished through a magnetic circuit consisting of a metallic core wrapped with primary and secondary windings of wire. The ratio of the number of primary windings to secondary windings is called the turn ratio. The voltage at the secondary can be increased or decreased depending on this ratio. In the ideal case, or lossless transformer, the product of secondary voltage and current will equal the product of primary voltage and current. That is, the ideal transformer does not dissipate power itself, but rather transforms power from one scenario to another. Real transformers dissipate some power because the copper wires have finite resistance and the magnetic coupling is not 100% efficient. The lost energy is often found in the form of heat. Another important characteristic of the transformer is that it creates electrical isolation between the primary and secondary. In other words, the circuit common points do not have to be the same potential or tied together between the primary and secondary sides.

Typically, power transformers are rated for a given input voltage and frequency (120 VAC/60 Hz in North America) which yields a specified secondary voltage under load. If the load current is minimal, the secondary voltage tends to increase beyond the rated value. This is due to the resistance of the windings and can be reduced by using a larger gauge although this results in a larger transformer. Also, it is common for secondary to be split or to have a center tap. A center tap allows the secondary to be treated as two symmetrical halves. This is useful for circuit rectification circuits. Finally, dots drawn on the transformer's schematic symbol and connections indicate like instantaneous polarity on the primary and secondary. That is, when the primary voltage is positive at its dot, the secondary voltage will also be positive at its dot.

10.3 Experimental Example
10.3.1 Equipment

(1) Dual channel oscilloscope model:_____ S/No.:_____
(1) Function generator model:_____ S/No.:_____
(1) DMM model:_____ S/No.:_____
(1) 12.6 volt, 1A center tapped transformer
(1) 10 Ω resistor ¼ watt actual: _____
(1) 22 Ω resistor ¼ watt actual: _____

(1) 20 Ω resistor 20 watt actual: _____

Schematic Diagrams

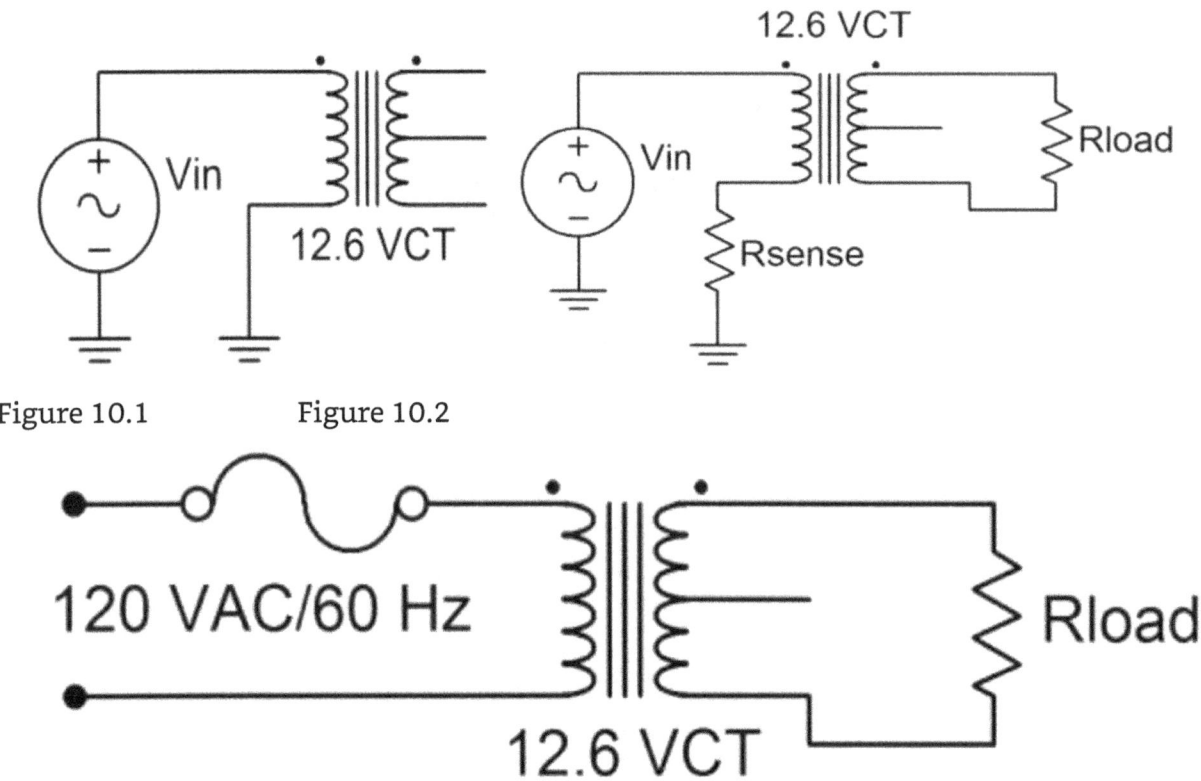

Figure 10.1   Figure 10.2

Figure 10.3

10.3.2 Procedure
Low Voltage
1. Consider the circuit of Figure 1. With a 12.6 volt secondary rating and a 120 volt primary rating, the turn ratio is approximately 10:1. In other words, for any reasonable input signal at the primary, the output at the secondary is expected to be one tenth the voltage and ten times the current.
2. Connect the primary side of the transformer to the function generator as shown in Figure 1. Set the generator to a 10 volt peak sine at 60 Hz. Place the oscilloscope probe grounds at the bottom of the secondary. Connect probe tip one to the top of the secondary and probe tip two to the center tap. Record the peak amplitudes in Table 10.1 and capture an image of the scope display. Compute and record the primary/secondary voltage ratio as well (for the full secondary).
3. Build the circuit of Figure 10.2 using R_{sense} = 22 Ω, R_{load} = 10 Ω and Vin = 5 volt peak sine at 60 Hz. Place one scope probe across the load and the other across R_{sense}. Record the peak amplitudes in Table 10.2 and capture an image of the scope display.
4. Using the voltage measured across the sense resistor, determine the primary side current. Using Ohm's law and the measured load voltage, determine the load (i.e., secondary) current. Based on these, compute the primary/secondary current ratio. Record these values in Table 10.2.

Line Voltage

5. This section uses the 120 VAC line. Treat it with the caution it deserves. Connect the circuit of Figure 10.3 leaving R_{load} unconnected. Measure the secondary voltage with the DMM (AC Volts). Record the value in Table 10.3 under "Unloaded".
6. Add the load resistor, 20 Ω, and measure the load voltage with the DMM. Record the value in Table 3 under "Loaded". Determine the percent change between the loaded and unloaded voltages. Also, measure the load voltage using the oscilloscope and capture an image of the display.

10.3.3 Result Tables

Primary Sense Voltage	
Primary Current	
Secondary Voltage	
Secondary Current	
Pri/Sec Current Ratio	

Table 10.1   Table 10.2

Full Secondary Voltage	
Center Tap Voltage	
Pri/Sec Voltage Ratio	

Table 10.3

V_{load} UnLoaded	
V_{load} Loaded	
Percent Change	

10.3.4 Questions

1. Examining the results of the circuit in Figure 1, does the specified turns ratio match that which is found experimentally? Why/why not?

_____
_____
_____
_____
_____

2. What is the effect of loading on a transformer's secondary voltage?

_____
_____
_____
_____
_____

3. Does the primary/secondary voltage ratio complement the primary/secondary current ratio? What does this say about the power dissipation of the transformer?

4. Are there appreciable variations between using the transformer at high input voltages versus low input voltages?

CHAPTER ELEVEN
Full-wave Bridge Rectifier

11.1 Objective
The objective of this exercise is to investigate the operation of a full-wave bridge rectifier as part of an AC to DC power supply. Also included are the effects of loading and filter capacitance.

11.2 Theory Overview
The full-wave bridge, like the half-wave rectifier, is used to turn an AC signal into pulsating DC. The fullwave bridge requires four diodes instead of one but has the advantage of utilizing the opposite polarity of the signal, effectively flipping its polarity rather than simply "throwing it away" like the half-wave circuit. This increases the energy available to the load and lessens the burden on filtering capacitors as the resulting gap between pulses is much smaller.

11.3 Experimental Example
11.3.1 Equipment
(1) Dual channel oscilloscope model:_____ S/No.:_____
(1) DMM model:_____ S/No.:_____
(4) Rectifying diodes (1N4002 series)
(1) 12.6 volt 1 amp center tapped transformer
(1) 1 k Ω resistor ¼ watt actual: _____
(1) 20 Ω resistor 20 watt actual: _____
(1) 1000 µF capacitor 25 volt actual: _____
1N4002 Datasheet: https://www.onsemi.com/pub/Collateral/1N4001-D.PDF

Schematic Diagrams

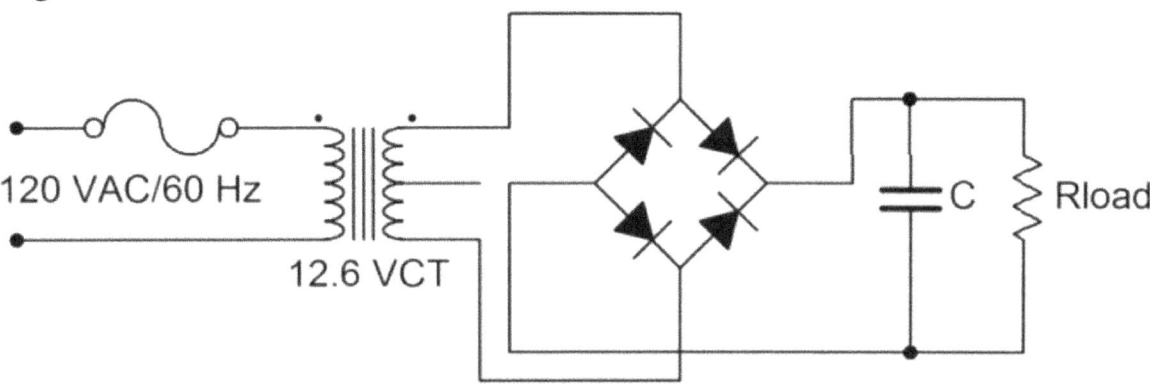

Figure 11.1

11.3.2 Procedure
Basic Operation
1. First, note that the circuit of Figure 11.1 is being powered directly from the AC line. Treat it with appropriate caution. It is worth repeating that any circuit should be de-

energized when making any changes to it.
2. Consider the circuit of Figure 11.1 without the capacitor connected. For a positive polarity of secondary voltage, the upper right and lower left diodes will be forward biased and allow current to flow through the load from top to bottom. The other two diodes will be reverse biased. For a negative secondary polarity the opposite occurs. That is, the upper left and lower right pair will be forward biased while the other two are reverse biased. This arrangement will also cause load current to flow through the load from top to bottom, thus effectively flipping the negative polarity portion of the wave.
3. Build the circuit of Figure 11.1 with R_{load} = 1 kΩ and C disconnected (open). This represents a very lightly loaded case. Under light loads, the output of the secondary will often be a little higher than the rated potential. Set the oscilloscope input to DC coupled. Measure and record the voltage across the secondary and then across the load. Do not use two probes to do this simultaneously as these two measurements do not share a common ground. Doing so will short out a portion of the circuit. Record the results in Table 11.1 and capture an image of the load voltage display.
4. Measure the load voltage with a DMM set to DC volts. Record this value in Table 11.1.
5. Replace the load with the 20 Ω resistors to simulate greater loading. Repeat steps 3 and 4.
6. Return the load resistor to the original 1 kΩ value and insert the 1000 μF capacitor. Measure the load voltage with both the oscilloscope and DMM, recording the values in Table 11.2. Be sure to capture an image of the scope display.
7. Replace the load with the 20 Ω resistors to simulate greater loading. Measure the load voltage with both the oscilloscope and DMM, recording the values in Table 2. Once again, be sure to capture an image of the scope display.

Computer Simulation
8. Simulate the circuit of Figure 11.1 using Transient Analysis. Use three variations, comparing the plotted waveforms to those measured in the laboratory: C = open with R_{load} = 20 Ω, C = 1000 μF with R_{load} = 1 kΩ, and C = 1000 μF with R_{load} = 20 Ω.

11.3.3 Result Tables
Table 11.1: No capacitor

Load	$V_{Secondary}$ Scope	V_{load} Scope	V_{load} DMM
1 kΩ			
20 Ω			

Table 11.2 with capacitor

Load	V_{load} Scope	V_{load} DMM
1 kΩ		
20 Ω		

11.3.4 Questions

1. What is the effect on the load voltage as the loading increases (i.e., as R_{load} decreases)?

2. What is the effect of adding the capacitor across the load?

3. How do the load voltages as measured by the DMM compare to those measured with the oscilloscope? Is there a pattern between the pairs of measurements?

4. How would the load voltages change if the diode bridge is connected between one end of the secondary and the center tap instead of across the entire secondary?

CHAPTER TWELVE
The DC Power Supply Project

12.1 Objective

This project involves the construction, testing and analysis of an Adjustable DC Power Supply. Each individual will complete a power supply and present a report to the lab section of the class. The presentation should be an overview of the report. The project will include the following:
1. Analysis of the circuit
2. Electronic Assembly of all components.
3. Testing, evaluation and troubleshooting.
4. Report Development
5. Report Presentation

12.2 Theory Overview

The circuit is an adjustable 5 to 12 V_{DC} supply utilizing a full-wave bridge and integrated voltage regulator. It picks up where the preceding Full-wave Rectifier exercise left off.

Schematic Diagram, and Artwork

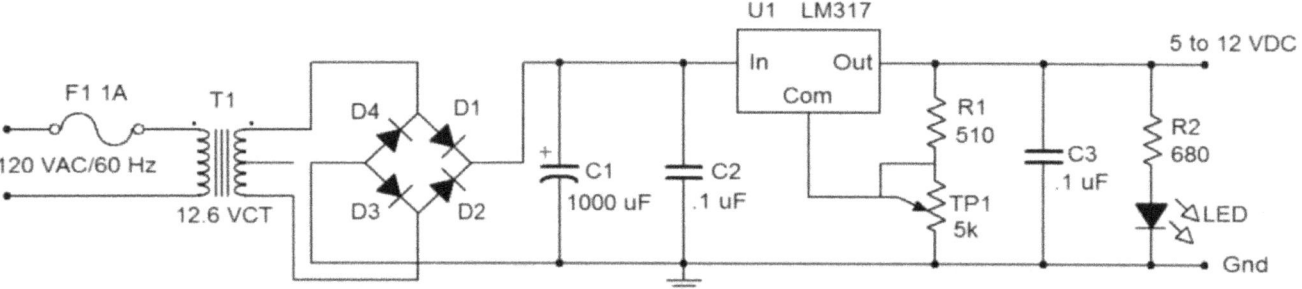

Figure 12.1

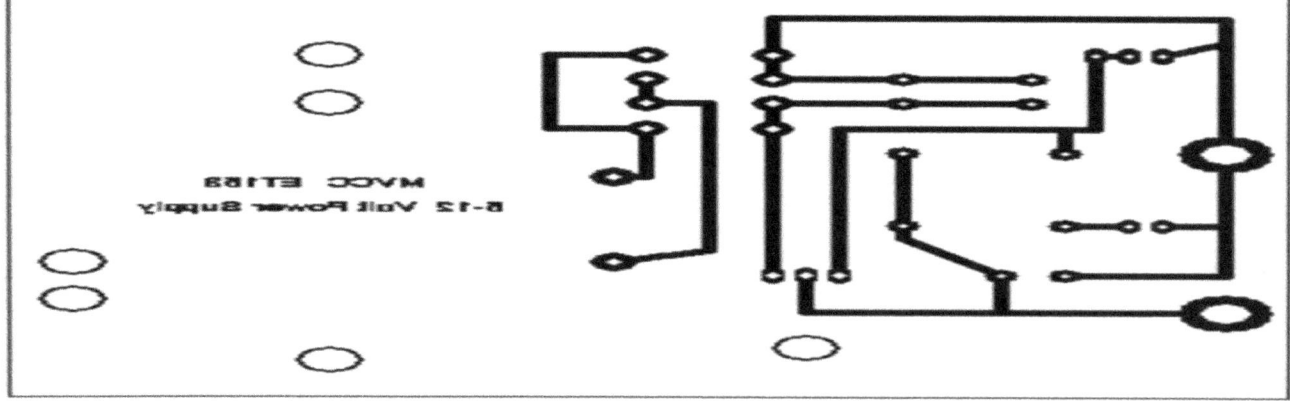

Figure 12.2 PCB Artwork (not actual size)

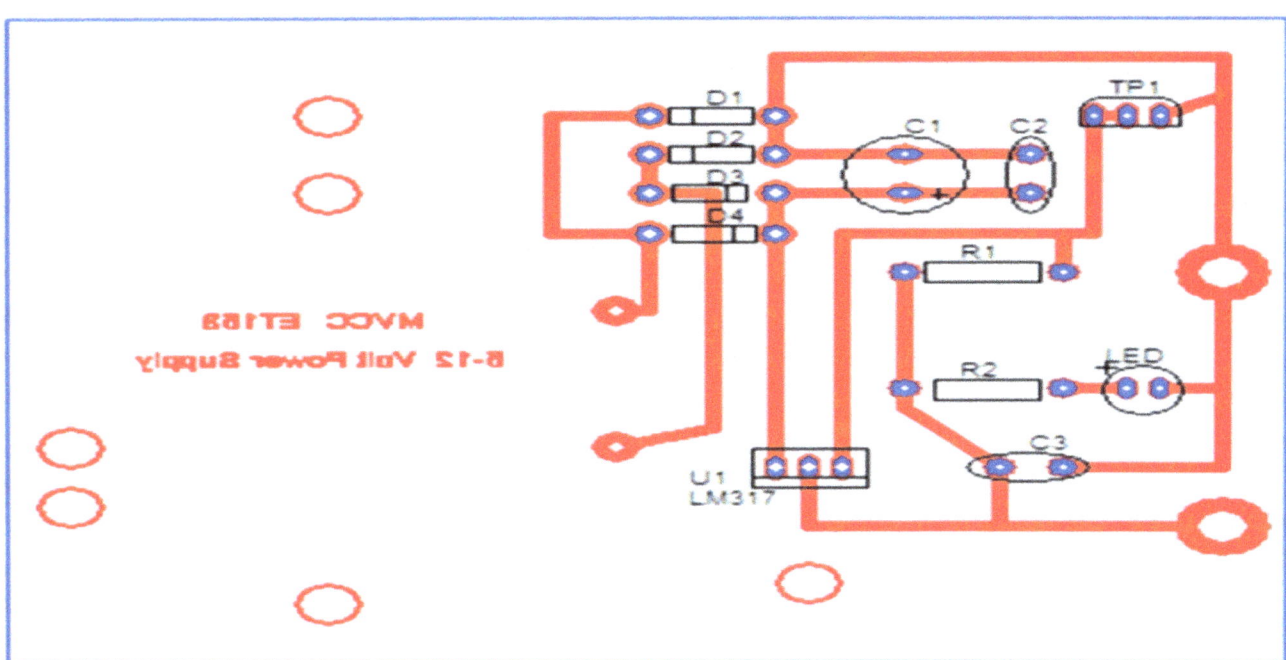

Figure 3, Component Placement

Table 12.1: 5 – 12 Volt DC Power Supply Bill of Materials

Item	Component ID	Description	Part No.	Qty
1	T1	117 - 12.6 Volt center tapped transformer		1
2	D1-D4	Rectifier diode	1N4001	4
3	C1	1000 microfarad 25 volt capacitor		1
4	C2, C3	0.1 microfarad 200 volt capacitor		2
5	R1	510 ohm, ± 5%, 0.5 watt		1
6	R2	680 ohm, ± 5%, 0.5 watt		1
7	TP1	5 k ohm trimpot, 10 turn ± 5%, 0.5 watt		1
8	LED	Red standard light emitting diode		1
9	U1	Voltage regulator	LM317T	1
10		Heat sink		1
11		Foot pads		4
12		Screws		5
13		Bolts		5
14		Washers		5
15		Wire tie		1
16		Line cord		1

| 17 | | In-line fuse holder and fuse | | 1 |

12.3 Report and Presentation Format

Report

The report should be in the following format:

Title Page – Includes Project Title, Course and Number, Name and Date

Description – Discuss the purpose and nature of the project.

Theory - Develop a block diagram for the circuit. Identify all the components that comprise each segment of the block diagram and describe their operation in terms of inputs and outputs. Include a listing of the specifications for the supply as follows:

Input power: 117 VAC RMS, ± 10%, 50 – 60 Hz

Output: 5 to 12 volts DC, adjustable, @ 300 mA with 0.5% ripple

Discuss the type of transformer used and describe the type of diode rectifier circuit. Also discuss the tolerance of the resistors in the circuit and the function of the potentiometer?

Equipment – List the special equipment that was used to complete the project.

Components – Refer to the PCB documentation. Include copies of all the printed circuit board documentation and discuss the purpose of each document.

Procedure – Describe the basic assembly and the general verification procedure for checking the power supply function, including circuit board assembly, soldering, and the testing procedures to verify key functions. Be sure to discuss the requirements for a good solder joint, the purpose of the flux in the solder and the need to frequently clean of the tip of the soldering iron. Also, identify the components that are polarity sensitive and discuss the impact of reversing the polarity during assembly.

Data – Record the measurements taken to verify the operation. These should include the Range of Adjustment, the No-Load Output Voltage, the Transformer Secondary Voltage, the Rectifier Voltage Output and the Peak to Peak Ripple at full output load.

Conclusion -What was learned on this project? What went wrong? How was it resolved? List any ideas about how to improve the performance of the power supply.

12.4 Oral Presentation

The presentation should be an overview of your report with a focus on your project conclusion. Your presentation will be video recorded.

Table 12.2: Report Grading Criteria

	1 = Below Minimum Standards	2 = At Minimum Standards	3 = At Average Standard	4 = Exceeds Average Standard
Format/Neatness 5%				
Grammar 5%				
Thoroughness 10%				

Theory Discussion 30%				
Procedure Discussion 30%				
Presentation 20%				

12.5 Circuit Verification Data Sheet

1. Output Voltage and LED Test: LED operational _____

Output Voltage Range of Adjustment _____ to _____

2. Adjust output to 5 V_{DC} with a 20 ohm load resistor attached to the output.

Calculated Load Current _____ Measured Load Current _____

Calculate the power being dissipated by the 20 ohm resistor? _____

3. Calculate the expected turn ratio of the transformer from the transformer data listed on the Bill of Material. Calculate and measure VSec RMS.

Calculated n = _____ Calculated VSec RMS = _____

Measured VPri RMS = _____ Measured VSec RMS = _____

4. Calculate the expected DC Voltage (V_{DC}) and Peak to Peak Ripple (Vrpp) out of the filter circuit using the formula: Vrpp = I_{LOAD}/fC where f = frequency and C is the value of the filter capacitor. Use a load current of 70 milliamps.

Calculated V_{DC} = _____ Calculated Vrpp = _____

5. Measure the DC Voltage (V_{DC}) and Peak to Peak Ripple (Vrpp) out of the filter circuit values with the 20 ohm resistor in place:

Measured V_{DC} = _____ Measured Vrpp = _____

6. Measure V_{DC} and Vrpp of the output voltage of the power supply with the 20 ohm load resistor in place.

VDC of the Power Supply Output = _____ Vrpp of the Power Supply Output = _____

CHAPTER THIRTEEN
Base Bias: CE Configuration

13.1 Objective

The objective of this exercise is to explore the operation of a basic common emitter biasing configuration for bipolar junction transistors, namely fixed base bias. Along with the general operation of the transistor and the circuit itself, circuit stability with changes in beta is also examined.

13.2 Theory Overview

For a bipolar junction transistor to operate properly, the base-emitter junction must be forward biased while the collector-base junction must be reverse biased. This will place VBE at approximately 0.7 volts and the collector current IC will be equal to the base current IB times the current gain β. For small signal devices, the current gain is greater than 100 typically. Thus, IC>>IB and IC≈IE.

The common emitter configuration places the emitter terminal at ground. The base terminal is seen as the input and the collector as the output. Using a fixed base supply, the base current is dependent on the value of the base resistor via Ohm's law. Consequently, any variation in current gain across a batch of transistors will show up as an equivalent variation in collector current, and by extension, a variation in collector-emitter voltage VCE.

13.3 Experimental Example

13.3.1 Equipment

(1) Dual adjustable DC power supply model:_____ S/No.:_____

(1) DMM model:_____ S/No.:_____

(3) Small signal NPN transistors (2N3904)

(1) 1.2 k Ω resistor ¼ watt actual: _____

(1) 330 k Ω resistor ¼ watt actual: _____

2N3904 Datasheet: https://www.onsemi.com/pub/Collateral/2N3903-D.PDF

Schematic Diagrams

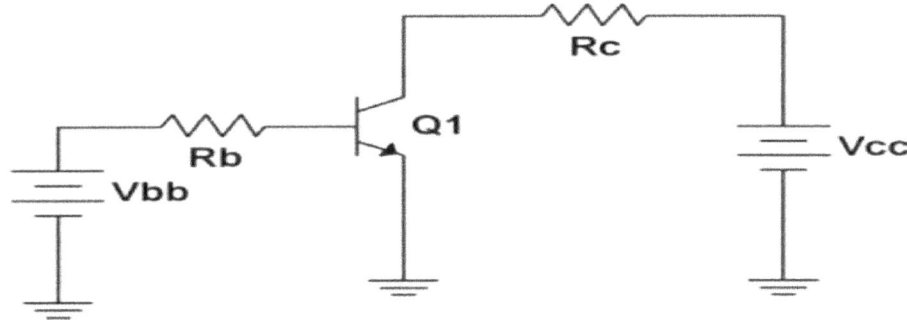

Figure 13.1

13.3.2 Procedure
A Quick Check
1. A quick and easy way to determine if a transistor is damaged is through the use of the resistance (or diode) function of a multimeter. The multimeter will produce a small current in order to determine the connected resistance value. This current is sufficient to partially forward or reverse bias a PN junction. Thus, for an NPN device, placing the red lead on the base and the black lead on the emitter and collector in turn will produce forward bias on the junctions and the meter will show a low -resistance. Reversing the leads will create reverse bias and a high resistance will be indicated. If the leads are connected from collector to emitter, one of the two junctions will be reverse biased regardless of lead polarity, and thus, a high resistance is always indicated. Before proceeding to the next step, check the three transistors using this method to ensure that they are functioning. (Note: some multimeters include a "beta checker" function. This may also be used to determine if the devices are good but the beta value produced should not be considered precise as the measurement current and voltage are most likely different from the circuit in which the transistor will be used.)

Base Bias

2. Consider the circuit of Figure 1 with Vbb = 11V, Vcc = 15V, Rb = 330 kΩ and Rc = 1.2 kΩ. Assume VBE = 0.7 volts. Further, assume that beta is 150 (a typical value for this device in this application). Calculate the expected values of IB, IC and IE, and record them in the "Theory" columns of Table 13.1.

Note that the theoretical values will be the same for all three transistors.

3. Based on the expected value of IC, determine the theoretical value of VCE and record it in Table 2. Also, fill in Table 13.2 with the typical (theoretical) beta value of 150.
4. Build the circuit of Figure 13.1 with Vbb = 11V, Vcc = 15V, Rb = 330 kΩ and Rc = 1.2 kΩ. Measure and record the base, collector and emitter currents, and record them in the first row of Table 13.1. Find the deviations between the theoretical and experimental currents, and record these in Table 13.1.
5. Measure the base-emitter and collector-emitter voltages and record in the first row of Table 12.2. Based on the measured values of base and collector current from Table 13.1, calculate and record the experimental betas in Table 13.2. Finally, compute and record the deviations for the voltages and for the current gain in Table 13.2.
6. Remove the first transistor and replace it with the second unit. Repeat steps four and five using the second row of Tables 13.1 and 2.
7. Remove the second transistor and replace it with the third unit. Repeat steps four and

five using the third row of Tables 13.1 and 2.

Design

8. One way of improving the circuit of Figure 13.1 is to redesign it so that a single power supply may be used. As noted previously, the base current is largely dependent on the value of VBB and RB. If the supply is changed, the resistance can be changed by a similar factor in order to keep the base current constant. This is just an application of Ohm's law. Based on this, determine a new value for RB that will produce the original IB if VBB is increased to the VCC value (i.e., a single power supply is used). Record this value in Table 13.3.

9. Rewire the circuit so that the original RB is replaced by the new calculated value (the nearest standard value will suffice). Also, the VBB supply should be removed and the left side of RB connected to the VCC supply. Measure the new base current and record it in Table 3. Also determine and record the deviation between the measured and target base current values.

Computer Simulation

10. Build the original circuit in a simulator. Run a single simulation and record the IB, IC, IE and VCE values in Table 13.4.

13.3.3 Result Tables

Table 13.1:

Transistor	$I_{B\ Theory}$	$I_{B\ Exp}$	$\%\ D\ I_B$	$I_{C\ Theory}$	$I_{C\ Exp}$	$\%\ D\ I_C$	$I_{E\ Theory}$	$I_{E\ Exp}$	$\%\ D\ I_E$
1									
2									
3									

Table 13.2:

Transistor	$V_{BE\ Theory}$	$V_{BE\ Exp}$	$\%\ D\ V_{BE}$	$V_{CE\ Theory}$	$V_{CE\ Exp}$	$\%\ D\ V_{CE}$	β_{Theory}	β_{Exp}	$\%\ D\ \beta$

r								
1								
2								
3								

Table 13.3:

Calculated R_B	Actual R_B Used	I_B Measured	% Deviation I_B

Table 13.4:

$I_{B\ sim}$	$I_{C\ sim}$	$I_{E\ sim}$	$V_{CE\ sim}$

13.3.4 Questions

1. Are the basic transistor parameters borne out in this exercise? That is, are the approximations of VBE = 0.7 and IC = IE valid?

   ___

2. Is the typical beta value of 150 highly accurate and repeatable?

   ___

3. Which circuit parameters are affected by beta changes? Which parameters appear to be immune to changes in beta?

   ___

4. Comparing Tables 1 and 2, is there a notable pattern between the deviations for beta and collector current? Why/why not?

   ___

5. In the circuit of Figure 1, what must RB be set to if VBB = 5V and the desired base current is 10 μA?

CHAPTER FOURTEEN
LED Driver Circuits

14.1 Objective
The objective of this exercise is to examine two methods of driving LEDs with a constant current. Method one involves a saturating switch while method two utilizes a non-saturating circuit.

14.2 Theory Overview
LEDs behave similarly to switching diodes in that they conduct current easily in forward bias and appear as an approximate open circuit when reverse biased. Unlike standard silicon switching diodes, however, the forward bias potential is not approximately 0.7 volts. Instead, this potential will vary depending on the design of the LED but typically will be in the neighborhood of 2 volts for everyday devices. The brightness of the LED is directly controlled by its current: the higher the current, the brighter the LED.
Consequently, it is important to drive LEDs with constant current sources to ensure consistent brightness.
Many circuits cannot drive LEDs directly so an intervening circuit is used (a driver) to boost the current up to the value the LED requires for a given brightness. The driving signal is attached to the base while the LED is situated in the collector, thus the transistor's current gain, beta, is exploited. Unfortunately, beta is not a particularly stable and consistent parameter so methods are required to alleviate this shortcoming. A saturating switch works by operating at the extreme ends of the DC load line; that is, either cutoff or saturation. If no signal is applied to the base, both base and collector currents will be zero, and thus, the LED is off. When a base signal is applied, it is designed to be large enough to force the transistor into saturation even with very modest betas. Therefore, the LED will always see the saturation current, regardless of the normal beta value. In contrast, the non-saturating circuit works by placing a resistor in the emitter. This establishes a constant emitter current (and thus, constant collector and LED current) in spite of beta changes. That is, if beta changes, the effect is seen in the base current, not the collector current. The non-saturating circuit has the advantage of using one less resistor, however, the saturating switch has the greater advantage of using the same collector and base voltages (the non-saturating circuit requires a collector source potential at least a few volts greater than the base voltage).

14.3 Experimental Example
14.3.1 Equipment
(1) Adjustable DC power supply model:_____ S/No.:_____
(1) DMM model:_____ S/No.:_____
(3) Small signal transistors (2N3904)
(1) LED
(1) 220 Ω resistor ¼ watt actual: _____
(1) 470 Ω resistor ¼ watt actual: _____

(1) 4.7 kΩ resistor ¼ watt actual: _____

Schematic Diagrams

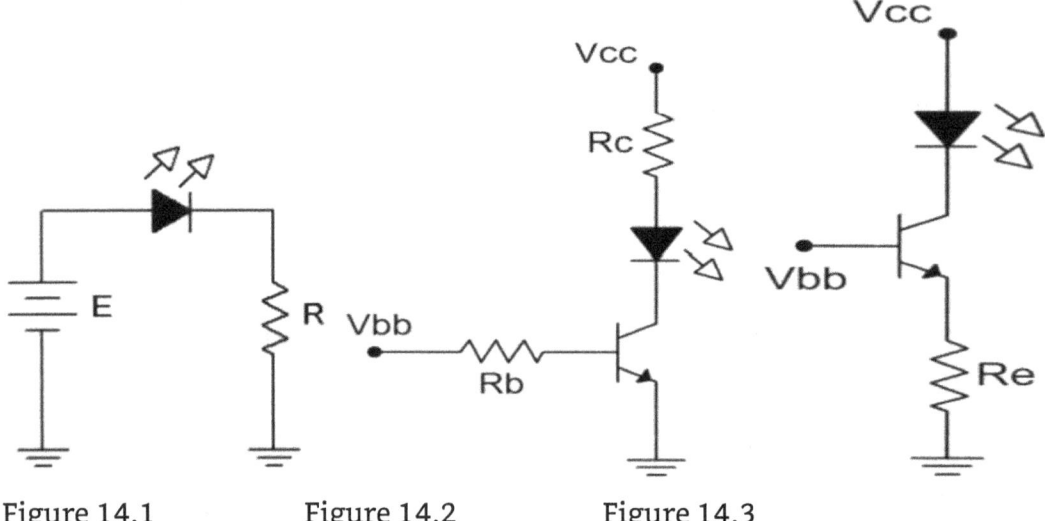

Figure 14.1　　　　Figure 14.2　　　　Figure 14.3

14.3.2 Procedure
Determining V_{LED}

1. The forward potential of an LED depends on its design and the current flowing through it. The other two circuits in this exercise are designed to produce LED currents of approximately 10 mA so a determination of the forward potential of this particular diode at 10 mA is desired. Assemble the circuit of Figure 14.1 using R = 470 Ω, and E = 5 volts. Insert an ammeter in line with the LED. Increase E until 10 mA is reached (the LED should be reasonably bright). Record the resulting LED voltage in Table 14.1.

Saturating Switch

2. Consider the saturating switch of Figure 14.2 using Vcc = V_{bb} = 5 volts, R_b = 4.7 kΩ and R_c = 220 Ω. Calculate the base and collector currents and record them in the first row of Table 14.2 (theory). As the circuit is in saturation, the theoretical V_{CE} is close to zero and may be found on the transistor data sheet via the V_{CE}/I_C saturation graph. Record this value in the first row of Table 14.2 as well.

3. Build the saturating switch of Figure 14.2 using V_{cc} = V_{bb} = 5 volts, R_b = 4.7 kΩ and R_c = 220 Ω. Measure and record the base and collector currents, and record the collector-emitter voltage in the first row of Table 14.2 (experimental). Also compute and record the deviations between theory and experimental results.

4. Remove the base resistor from V_{bb} and connect it to ground. Without a base source potential, the circuit will be in cutoff. Determine the theoretical base and collector currents along with the collectoremitter voltage and record them in the second row of Table 14.2. Measure these parameters, record them in Table 14.2, and also compute and record the resulting deviations.

5. Reconnect the base resistor to the V_{bb} supply and swap in the second transistor. Repeat steps 3 and 4 using the next two rows of Table 14.2.

6. Reconnect the base resistor to the V_{bb} supply and swap in the third transistor. Repeat steps 3 and 4 using the final two rows of Table 14.2.

Non-saturating Current Source

ELECTRONICS LABORATORY MANUAL

7. Consider the non-saturating current source of Figure 14.3 using V_{cc} = 10 volts, V_{bb} = 5 volts and R_e = 470 Ω. Using a typical beta of 150, calculate the base and collector currents, and the collectoremitter voltage and record them in the first row of Table 14.3 (theory).
8. Build the non-saturating current source of Figure 14.3 using V_{cc} = 10 volts, V_{bb} = 5 volts and Re = 470 Ω. Measure and record the base and collector currents, and record the collector-emitter voltage in the first row of Table 14.3 (experimental). Also compute and record the deviations between theory and experimental results.
9. Remove V_{bb} and connect the base terminal to ground. Without the base source potential, the base current will be zero. Determine the theoretical base and collector currents along with the collector emitter voltage and record them in the second row of Table 14.3. Measure these parameters, record them in Table 14.3, and also compute and record the resulting deviations.
10. Reconnect the V_{bb} supply to the base and swap in the second transistor. Repeat steps 8 and 9 using the next two rows of Table 14.3.
11. Reconnect the base resistor to the V_{bb} supply and swap in the third transistor. Repeat steps 8 and 9 using the final two rows of Table 14.3.

Design

12. As seen in steps 7 through 11, the LED current of Figure 14.3 is a function of the base supply and the emitter resistor. Determine a new value for the emitter resistance that will yield an LED current of 15 mA. Record this value in Table 14.4. Obtain a new resistor close in value to the calculated result and swap it into the circuit. Measure the resulting LED current and record in Table 14.4.

Computer Simulation

13. Simulate the circuit of Figure 14.2 and record the currents and voltage in Table 14.5.
14. Simulate the circuit of Figure 14.3 and record the currents and voltage in Table 14.6.

14.3.3 Result Tables

Table 14.1:

V_{LED}	

Table 14.2

V_{bb}	$I_{B\,Theory}$	$I_{C\,Theory}$	$V_{CE\,Theory}$	$I_{B\,Exp}$	$I_{C\,Exp}$	$V_{CE\,Exp}$	% DI_B	% DI_C	% DV_{CE}
5									
0									
5									
0									
5									
0									

Table 14.3

V_{bb}	$I_{B\,Theory}$	$I_{C\,Theory}$	$V_{CE\,Theory}$	$I_{B\,Exp}$	$I_{C\,Exp}$	$V_{CE\,Exp}$	% DI_B	% DI_C	% DV_{CE}
5									

0									
5									
0									
5									
0									

Table 14.4

V_{bb}	$R_{E\,Theory}$	$R_{E\,Exp}$	$I_{C\,Exp}$

Table 14.5

V_{bb}	$I_{B\,Sim}$	$I_{C\,Sim}$	$V_{CE\,Sim}$
5			
0			

Table 14.6

V_{bb}	$I_{B\,Sim}$	$I_{C\,Sim}$	$V_{CE\,Sim}$
5			
0			

14.3.4 Questions

1. Do the two driver circuits produce a stable and predictable LED current in spite of changes in beta?

   _____
   _____
   _____
   _____
   _____

2. The circuit of Figure 2 is stated to be a saturating switch. How do the data of Table 2 confirm this statement?

   _____
   _____
   _____
   _____
   _____

3. The circuit of Figure 3 is stated to be a non-saturating current source. How do the data of Table 3 confirm this statement?

   _____
   _____

_____
_____
_____

CHAPTER FIFTEEN
Emitter Bias

15.1 Objective
The objective of this exercise is to examine the two supply emitter bias topology and determine whether or not it produces a stable Q point. Various potential troubleshooting issues are also explored.

15.2 Theory Overview
One of the problems with simpler biasing schemes such as the base bias is that the Q point (IC and VCE) will fluctuate with changes in beta. This will result in inconsistent circuit performance. If a fixed voltage can be developed across an emitter resistor, a stable Q point will result. To obtain this fixed emitter resistor voltage, a negative voltage supply may be connected to the low side of the emitter resistor instead of connecting that resistor to ground. The transistor's base is then simply connected back to ground via a single resistor. If this base resistance is relatively small, the base voltage will be close to zero as only the base current flows through it. Consequently, almost all of the negative emitter supply will drop across the emitter resistor, with the exception of the single base-emitter potential. This will result in a stable emitter current, and by extension, stable collector current and collector-emitter voltage. As beta varies, this change will be reflected in a change in base current. This can result in large percentage changes in base voltage; however, the magnitude of the base potential will remain small, and thus, inconsequential.

15.3 Experimental Example
15.3.1 Equipment
(1) Dual adjustable DC power supply model:_____ S/No.:_____
(1) DMM model:_____ S/No.:_____
(3) Small signal transistors (2N3904)
(1) 15 k Ω resistor ¼ watt actual: _____
(1) 22 k Ω resistor ¼ watt actual: _____
(1) 33 k Ω resistor ¼ watt actual: _____

Schematic Diagram

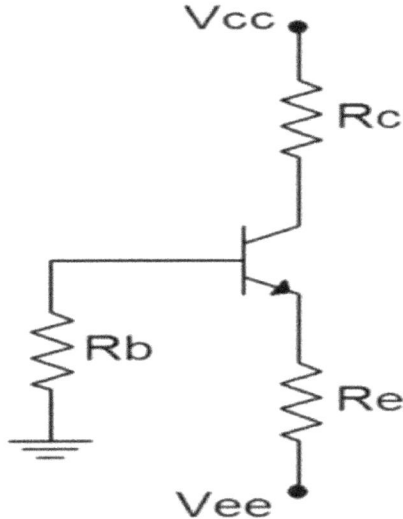

Figure 15.1

15.3.2 Procedure
DC Load Line

11. Consider the circuit of Figure 15.1 using Vcc = 15 volts, Vee = -12 volts, Rb = 33 kΩ, Re = 22 kΩ and Rc = 15 kΩ. Using the approximation of a negligible base voltage, determine the ideal end points of the DC load line and the Q point, and record these in Table 15.1.

Circuit Voltages and Beta

12. Continuing with the component values indicated in step one, compute the theoretical emitter and collector voltages, and record them in Table 15.2 (Theory). For the theoretical base voltage entry, assume a beta of approximately 150 and determine the base current and voltage from the theoretical collector current recorded in the Table 15.1.

13. Build the circuit of Figure 15.1 using Vcc = 15 volts, Vee = -12 volts, Rb = 33 kΩ, Re = 22 kΩ and Rc = 15 kΩ. Measure the base, emitter and collector voltages and record them in the first row of Table 15.2 (Experimental). Compute the deviations between theoretical and experimental and record these in the first row of Table 15.2 (% Deviation).

14. Measure the base and collector currents and record these in the first row of Table 15.3. Based on these, compute and record the experimental beta as well.

15. Swap the transistor with the second transistor and repeat steps 3 and 4 using the second rows of the tables.

16. Swap the transistor with the third transistor and repeat steps 3 and 4 using the third rows of the tables.

Design

17. The collector voltage of the circuit can be altered by a variety of means including changing the collector resistance. If the emitter supply and resistance are held constant, the collector voltage is determined by the collector resistance and the collector supply. Redesign the circuit to achieve a collector voltage of approximately 10 volts. Obtain a resistor close to this value, swap out the original collector resistor and measure the resulting voltage. Record the appropriate values in Table 4.

Troubleshooting

18. Return the original collector resistor to the circuit. Consider each of the individual faults listed in Table 15.5 and estimate the resulting base, emitter and collector voltages. Introduce each of the individual faults in turn and measure and record the transistor voltages in Table 15.5.

Computer Simulation

19. Build the circuit of Figure 1 in a simulator. Run a DC simulation and record the resulting transistor voltages in Table 15.6.

15.3.3 Result Tables

Table 15.1

$V_{CE\,(Cutoff)}$	
$I_{C\,(sat)}$	
V_{CEQ}	
I_{CQ}	

Table 15.2

Transistor	$V_{B\,Theory}$	$V_{E\,Theory}$	$V_{C\,Theory}$	$V_{B\,Exp}$	$V_{E\,Exp}$	$V_{C\,Exp}$	$\%\,DV_B$	$\%\,DV_E$	$\%\,DV_C$
1									
2									
3									

Table 15.3

Transistor	I_B	I_C	β
1			
2			
3			

Table 15.4

$R_{C\,Theory}$	$R_{C\,Actual}$	$V_{C\,Measured}$

Table 15.5

Issue	V_B	V_E	V_C
R_B Short			
R_B Open			
R_C Short			
R_C Open			
R_B Open			

V_{CB} Open			

Table 15.6

V_B Sim	
V_E Sim	
V_C Sim	

15.3.4 Questions

1. Based on the results of Table 15.1, is the transistor operating in saturation, cutoff or in the linear region?

2. Based on the results of Tables 15.2 and 15.3, does the circuit achieve a stable operating point when compared to beta?

3. How does the Emitter Bias circuit compare to Base Bias in terms of Q point stability and complexity?

4. Using the original circuit, determine a new value for the emitter resistance that will yield half of the quiescent collector current recorded in Table 15.1.

CHAPTER SIXTEEN
Voltage Divider Bias

16.1 Objective

The objective of this exercise is to examine the voltage divider bias topology and determine whether or not it produces a stable Q point. Various potential troubleshooting issues are also explored.

16.2 Theory Overview

Like Emitter Bias, Voltage Divider Bias seeks to establish a stable Q point by placing a fixed voltage across an emitter resistor. This will result in a stable emitter current, and by extension, stable collector current and collector-emitter voltage. As beta varies, this change will be reflected in a change in base current. With proper design, this change in base current will have little overall impact on circuit performance. One method of obtaining a stable voltage across the emitter resistor is to apply a stiff voltage divider to the base. "Stiff", in this case, means that the current through the divider resistors should be much higher than the current tapped off of the divider (the current being tapped off is the base current).

By doing so, variations in base current will not excessively load the divider and this will lead to a very stable base voltage. The emitter voltage is one base-emitter drop less, and is the potential across the emitter resistor. Hence, the emitter resistor's voltage will be kept stable.

When troubleshooting, circuit faults often result in either shorted or open components. Typically this will alter the circuit radically and push the Q point into either cutoff or saturation. The fault may also alter the DC load line itself. Once the transistor goes into either cutoff or saturation, normal linear operation will be lost.

16.3 Experimental Example
16.3.1 Equipment
(1) Adjustable DC Power Supply model:_____ S/No.:_____

(1) DMM model:_____ S/No.:_____

(3) Small signal transistors (2N3904)

(1) 3.3 k Ω resistor ¼ watt actual: _____

(1) 4.7 k Ω resistor ¼ watt actual: _____

(1) 5.6 k Ω resistor ¼ watt actual: _____

(1) 10 k Ω resistor ¼ watt actual: _____

Schematic Diagram

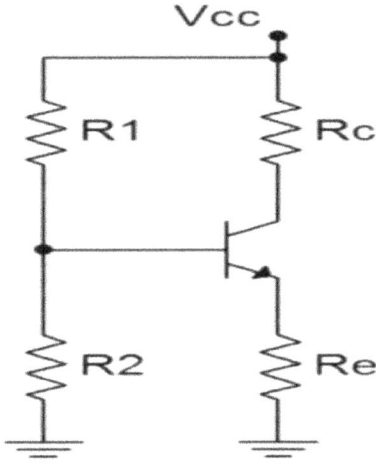

Figure 16.1

16.3.2 Procedure

DC Load Line

1. Consider the circuit of Figure 16.1 using Vcc = 10 volts, R1 = 10 kΩ, R2 = 3.3 kΩ, Re = 4.7 kΩ and Rc = 5.6 kΩ. Using the approximation of a lightly loaded "stiff" voltage divider, determine the ideal end points of the DC load line and the Q point, and record these in Table 16.1.

Circuit Voltages and Beta

2. Continuing with the component values indicated in step one, compute the theoretical base, emitter and collector voltages, and record them in Table 16.2 (Theory).
3. Build the circuit of Figure 16.1 using Vcc = 10 volts, R1 = 10 kΩ, R2 = 3.3 kΩ, Re = 4.7 kΩ and Rc = 5.6 kΩ. Measure the base, emitter and collector voltages and record them in the first row of Table 16.2 (Experimental). Compute the deviations between theoretical and experimental and record these in the first row of Table 16.2 (% Deviation).
4. Measure the base and collector currents and record these in the first row of Table 16.3. Based on these, compute and record the experimental beta as well.
5. Swap the transistor with the second transistor and repeat steps 3 and 4 using the second rows of the tables.
6. Swap the transistor with the third transistor and repeat steps 3 and 4 using the third rows of the tables.

Design

7. The collector current of the circuit can be altered by a variety of means including changing the emitter resistance. If the base voltage is held constant, the collector current is determined by the emitter resistance via Ohm's law. Redesign the circuit to achieve half of the quiescent collector current recorded in Table 16.1. Obtain a resistor close to this value, swap out the original emitter resistor and measure the resulting current. Record the appropriate values in Table 16.4.

Troubleshooting

8. Return the original emitter resistor to the circuit. Consider each of the individual faults listed in Table 16.5, and estimate the resulting base, emitter and collector voltages. Introduce each of the individual faults in turn and measure and record the transistor voltages in Table 16.5.

Computer Simulation

9. Build the circuit of Figure 16.1 in a simulator. Run a DC simulation and record the resulting transistor voltages in Table 16.6.

16.3.3 Result Tables

Table 16.1

	$V_{CE\,(Cutoff)}$	
	$I_{C\,(sat)}$	
	V_{CEQ}	
	I_{CQ}	

Table 16.2

Transistor	$V_{B\,Theory}$	$V_{E\,Theory}$	$V_{C\,Theory}$	$V_{B\,Exp}$	$V_{E\,Exp}$	$V_{C\,Exp}$	%DV_B	%DV_E	%DV_C
1									
2									
3									

Table 16.3

Transistor	I_B	I_C	β
1			
2			
3			

Table 16.4

$R_{C\,Theory}$	$R_{C\,Actual}$	$V_{C\,Measured}$

Table 16.5

Issue	V_B	V_E	V_C
R_B Short			
R_B Open			

R_C Short				
R_C Open				
R_B Open				
V_{CB} Open				

Table 16.6

	V_B Sim	
	V_E Sim	
	V_C Sim	

16.3.4 Questions

1. Based on the results of Table 16.1, is the transistor operating in saturation, cutoff or in the linear region?

   _____
   _____
   _____
   _____
   _____

2. Based on the results of Tables 16.2 and 3, does the circuit achieve a stable operating point when compared to beta?

   _____
   _____
   _____
   _____
   _____

4. Based on the measurements of Table 16.5, is it possible for different circuit problems to produce similar or even identical voltages in the circuit, or is every fault unique in its outcome?

   _____
   _____
   _____
   _____
   _____

5. What is the required design condition for the voltage divider bias to achieve high Q point stability in spite of changes in beta?

   _____
   _____
   _____

6. Using the original circuit, determine a new value for the collector resistance that will yield a collector voltage of approximately half of the power supply value.

CHAPTER SEVENTEEN
Feedback Biasing

17.1 Objective

The objective of this exercise is to examine two kinds of feedback biasing: collector feedback and emitter feedback. Both forms potentially are more stable than simple base bias in terms of the impact of beta on collector current.

17.2 Theory Overview

By inserting a resistor in either the emitter or collector portions of the transistor circuit, it is possible to partially control the base current in such a way that an increase in beta will cause a decrease in base current which in turn helps to mitigate the tendency of collector current to increase. This will result in circuits that have greater Q point stability than simple base bias circuits although for certain practical reasons they might not be as stable as voltage divider or dual supply emitter bias schemes.

In the collector feedback arrangement, the base resistor is connected from the collector to the base. Therefore, its voltage is one base-emitter drop less than the collector voltage. The collector voltage, in turn, is simply the supply potential minus the collector resistor's drop. Therefore, as the collector current rises, the collector resistor's drop increases, forcing the collector voltage down and thus reducing the base resistor's voltage. By Ohm's law, this means that the base current must decrease. This decrease helps to limit the overall increase in collector current.

The emitter feedback situation is similar. In this instance, as collector current increases the drop across the emitter resistor rises. This will result in an increase in base voltage as it is locked to one base-emitter drop above the emitter. Consequently, as the collector current increases, the voltage across the base resistor decreases which helps to compensate for the original increase in collector current.

17.3 Experimental Example
17.3.1 Equipment

(1) Adjustable DC Power Supply model:_____ S/No.:_____

(1) DMM model:_____ S/No.:_____

(3) Small signal transistors (2N3904)

(1) 330 Ω resistor ¼ watt actual: _____

(1) 470 Ω resistor ¼ watt actual: _____

(1) 1 k Ω resistor ¼ watt actual: _____

(1) 220 k Ω resistor ¼ watt actual: _____

Schematic Diagrams

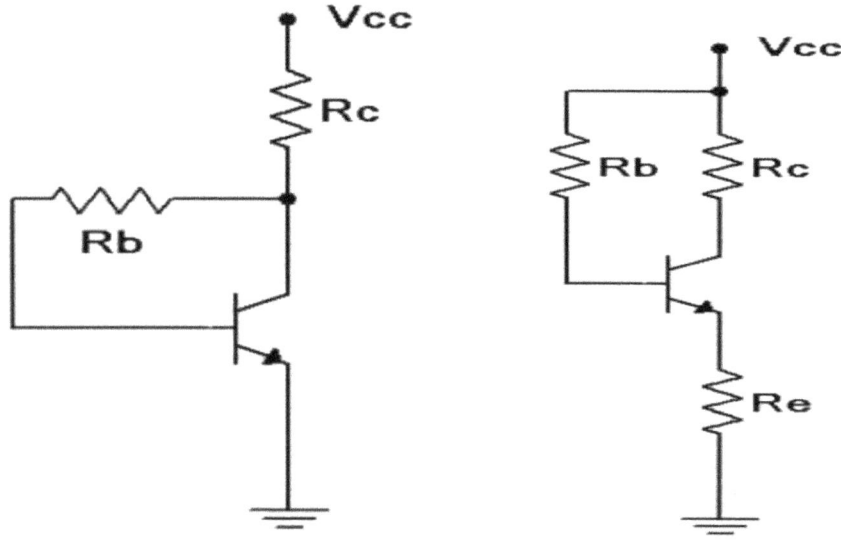

Figure 17.1   Figure 17.2

17.3.2 Procedure

Collector Feedback - DC Load Line

1. Consider the circuit of Figure 17.1 using Vcc = 12 volts, Rb = 220 kΩ and Rc = 1 kΩ. Determine the ideal end points of the DC load line and the Q point, and record these in Table 17.1.

Circuit Voltages and Beta

2. Continuing with the component values indicated in step one, compute the theoretical base, emitter and collector voltages, and record them in Table 17.2 (Theory).
3. Build the circuit of Figure 17.1 using Vcc = 12 volts, Rb = 220 kΩ and Rc = 1 kΩ. Measure the base, emitter and collector voltages and record them in the first row of Table 17.2 (Experimental).
4. Measure the base and collector currents and record these in the first row of Table 17.3. Based on these, compute and record the experimental beta as well.
5. Swap the transistor with the second transistor and repeat steps 3 and 4 using the second rows of the tables.
6. Swap the transistor with the third transistor and repeat steps 3 and 4 using the third rows of the tables.

Stability

7. Based on the measurements of Table 17.3, determine the maximum percent change of collector current and record in Table 17.4. Also determine the maximum percent change of beta and record in Table 17.4.

Emitter Feedback - DC Load Line

8. Consider the circuit of Figure 17.2 using Vcc = 12 volts, Rb = 220 kΩ, Re = 470 Ω and Rc = 330 Ω. Determine the ideal end points of the DC load line and the Q point, and record these in Table 17.5.

Circuit Voltages and Beta

9. Continuing with the component values indicated in step one, compute the theoretical base, emitter and collector voltages, and record them in Table 17.6 (Theory).
10. Build the circuit of Figure 17.2 using Vcc = 12 volts, Rb = 220 kΩ, Re = 470 Ω and Rc = 330 Ω. Measure the base, emitter and collector voltages and record them in the first row of Table 17.6 (Experimental).
11. Measure the base and collector currents and record these in the first row of Table 17.7. Based on these, compute and record the experimental beta as well.
12. Swap the transistor with the second transistor and repeat steps 3 and 4 using the second rows of the tables.
13. Swap the transistor with the third transistor and repeat steps 3 and 4 using the third rows of the tables.

Stability

14. Based on the measurements of Table 17.7, determine the maximum percent change of collector current and record in Table 17.8. Also determine the maximum percent change of beta and record in Table 17.8.

Troubleshooting

15. For the emitter feedback bias circuit, consider each of the individual faults listed in Table 17.9 and estimate the resulting base, emitter and collector voltages. Introduce each of the individual faults in turn and measure and record the transistor voltages in Table 17.9.

17.3.3 Result Tables

Table 17.1

	$V_{CE\,(Cutoff)}$	
	$I_{C\,(sat)}$	
	V_{CEQ}	
	I_{CQ}	

Table 17.2

Transistor	$V_{B\,Theory}$	$V_{E\,Theory}$	$V_{C\,Theory}$	$V_{B\,Exp}$	$V_{E\,Exp}$	$V_{C\,Exp}$	% DV_B	% DV_E	% DV_C
1									
2									
3									

Table 17.3

Transistor	I_B	I_C	β
1			
2			
3			

Table 17.4

	%Δβ	
	%ΔI_C	

Table 17.5

	$V_{CE(Cutoff)}$	
	$I_{C(sat)}$	
	V_{CEQ}	
	I_{CQ}	

Table 17.6

Transistor	$V_{B\,Theory}$	$V_{E\,Theory}$	$V_{C\,Theory}$	$V_{B\,Exp}$	$V_{E\,Exp}$	$V_{C\,Exp}$	%DV_B	%DV_E	%DV_C
1									
2									
3									

Table 17.7

Transistor	I_B	I_C	β
1			
2			
3			

Table 17.8

	%Δβ	
	%ΔI_C	

Table 17.9

Issue	V_B	V_E	V_C
R_B Short			
R_B Open			
R_C Short			
R_C Open			
R_B Open			
V_{CB} Open			

17.3.4 Questions

1. Based on the results of Tables 4 and 8, do these circuits achieve a stable operating point when compared to beta?

   _____
   _____
   _____
   _____
   _____

2. Which circuit is more stable in this exercise; the emitter feedback bias or the collector feedback bias? Is it safe to say that this will always be the case for any emitter feedback bias circuit versus any collector feedback bias circuit?

   _____
   _____
   _____
   _____
   _____

3. Are the transistor voltages always as stable as the collector current or can they be more or less stable?

   _____
   _____
   _____
   _____
   _____

4. Based on the collector current equation derivations for the two circuits, derive the collector current equation for a combination circuit which consists of a collector feedback bias circuit with an emitter resistor added.

   _____
   _____
   _____
   _____
   _____

CHAPTER EIGHTEEN
PNP Transistors

18.1 Objective

The objective of this exercise is to investigate the practical differences between circuits implemented with PNP transistors versus NPN transistors. PNP versions of basic biasing and LED driver circuits will be used.

18.2 Theory Overview

On a practical level, PNP transistors may be thought of as a mirror image of their NPN counterparts. That is, all of the device's voltage polarities and current directions will be opposite of those found with NPNs.

In fact, a simple way to turn an NPN circuit into an equivalent PNP circuit is to swap out the transistor and then flip the polarity of the power supply (or supplies, as the case may be). The resulting circuit will produce essentially the same voltages and currents as the original but with reversed polarities. By no means are negative power supplies a requirement to use PNPs, though. Commonly, the circuit is "flipped top to bottom" and implemented with a positive supply. In this case the emitter will be found toward the top and the collector toward the bottom. In some instances this orientation may also reverse the operational logic of the circuit. For example, the "flipped" PNP LED driver becomes an inverting driver.

That is, a logic low will light the LED instead of logic high.

18.3 Experimental Example
18.3.1 Equipment

(1) Adjustable DC Power Supply model:_____ S/No.:_____

(1) DMM model:_____ S/No.:_____

(3) Small signal PNP transistors (2N3906)

(1) LED

(1) 220 Ω resistor ¼ watt actual: _____

(1) 3.3 kΩ resistor ¼ watt actual: _____

(1) 4.7 kΩ resistor ¼ watt actual: _____

(1) 5.6 kΩ resistor ¼ watt actual: _____

(1) 10 kΩ resistor ¼ watt actual: _____

2N3906 Datasheet: https://www.onsemi.com/pub/Collateral/2N3906-D.PDF

Schematic Diagram

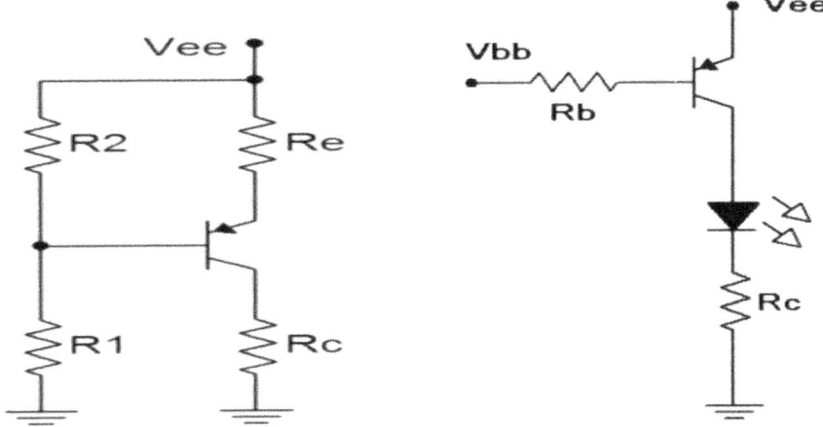

Figure 18.1   Figure 18.2

18.3.2 Procedure

PNP Voltage Divider

1. Consider the circuit of Figure 18.1 using Vee = 10 volts, R1 = 10 kΩ, R2 = 3.3 kΩ, Re = 4.7 kΩ and Rc = 5.6 kΩ. Using the approximation of a lightly loaded "stiff" voltage divider, determine the theoretical base, emitter and collector voltages, and record them in Table 18.1 (Theory).
2. Build the circuit of Figure 18.1 using Vee = 10 volts, R1 = 10 kΩ, R2 = 3.3 kΩ, Re = 4.7 kΩ and Rc = 5.6 kΩ. Measure the base, emitter and collector voltages and record them in the first row of Table 18.1 (Experimental).
3. Swap the transistor with the second transistor and repeat steps 1 and 2 using the second row of the table.
4. Swap the transistor with the third transistor and repeat steps 1 and 2 using the third row of the table.

Troubleshooting

5. Consider each of the individual faults listed in Table 18.2 and estimate the resulting base, emitter and collector voltages. Introduce each of the individual faults in turn and measure and record the transistor voltages in Table18.2.

PNP LED Driver

6. Consider the PNP saturating switch of Figure 18.2 using Vee = Vbb = 5 volts, Rb = 4.7 kΩ and Rc = 220 Ω. Calculate the base and collector currents and record them in the first row of Table 18.3 (Theory). As the circuit is in saturation, the theoretical VCE is close to zero and may be found on the transistor data sheet via the VCE/IC saturation graph. Record this value in the first row of Table 18.3 as well.
7. Build the saturating switch of Figure 18.2 using Vee = Vbb = 5 volts, Rb = 4.7 kΩ and Rc = 220 Ω. Measure and record the base and collector currents, and record the collector-emitter voltage in the first row of Table 18.3 (Experimental). Also compute and record the deviations between theory and experimental results.
8. Remove the base resistor from Vbb and connect it to ground. Without a base source potential, the circuit will be in cutoff. Determine the theoretical base and collector currents along with the collector emitter voltage and record them in the second row

of Table 18.3. Measure these parameters, record them in Table 18.3, and also compute and record the resulting deviations.
9. Reconnect the base resistor to the Vbb supply and swap in the second transistor. Repeat steps 3 and 4 using the next two rows of Table 18.3.
10. Reconnect the base resistor to the Vbb supply and swap in the third transistor. Repeat steps 3 and 4 using the final two rows of Table 18.3.

Design

11. A simple way to program the LED current in the driver is by altering the collector resistor. First, measure the LED potential while it is lit. Assuming that the collector-emitter saturation voltage is negligible, all of the power supply voltage will drop across the collector resistor when the LED is lit, with the exception of the LED voltage. Ohm's law can then be used to determine a resistance value for a desired target current. Compute the required value of resistance to achieve an LED current of 8 mA. Replace the collector resistor with the nearest value available and measure the resulting current. Record the appropriate values in Table 18.4.

18.3.3 Result Tables

Table 18.1

Transistor	$V_{B\ Theory}$	$V_{E\ Theory}$	$V_{C\ Theory}$	$V_{B\ Exp}$	$V_{E\ Exp}$	$V_{C\ Exp}$
1						
2						
3						

Table 18.2

Issue	V_B	V_E	V_C
R_2 Short			
R_E Open			
R_C Short			
R_C Open			
V_{CE} Short			
V_{CE} Open			

Table 18.3

V_{bb}	$I_{B\ Theory}$	$I_{C\ Theory}$	$V_{CE\ Theory}$	$I_{B\ Exp}$	$I_{C\ Exp}$	$V_{CE\ Exp}$
5						
0						

ELECTRONICS LABORATORY MANUAL

5					
0					
5					
0					

Table 18.4

$R_{C\,Theory}$	$R_{C\,Actual}$	$I_{C\,Measured}$

18.3.4 Questions

1. Is the PNP voltage divider circuit as stable as its NPN counterpart studied earlier?

2. Compare the NPN voltage divider lab results to this PNP version. If the various transistor voltages are added together (e.g., NPN base voltage plus PNP base voltage), a constant results. What is the significance of this value and will it always work out in this fashion? Why/why not?

3. Do the troubleshooting faults presented in the PNP circuit produce similar transistor voltages compared to the same faults in the NPN version of the circuit? Why/why not?

4. How does the operational logic of the PNP LED driver compare to the NPN version of the same circuit?

5. Are the LED current design considerations the same as those of the NPN version?

CHAPTER NINETEEN
Common Emitter Amplifier

19.1 Objective

The objective of this exercise is to examine the characteristics of a common emitter amplifier, specifically voltage gain, input impedance and output impedance. A method for experimentally determining input and output impedance is investigated along with various potential troubleshooting issues.

19.2 Theory Overview

An ideal common emitter amplifier simply multiples the input function by a constant value while also inverting the signal. The voltage amplification factor, Av, is largely a function of the AC load resistance at the collector and the internal emitter resistance, r'_e. This internal resistance is, in turn, inversely proportional to the DC emitter current. Therefore, if the underlying bias is stable with changes in beta, the voltage gain will also be stable. The circuit will appear as impedance to the signal source, Z_{in}. This impedance is approximately equal to the base biasing resistor(s) in parallel with the impedance seen looking into the base (Z_{in}(base)) which is approximately equal to $\beta \, r'_e$. Consequently, the amplifier's input impedance may experience some variation with beta. In contrast, the circuit's output impedance as seen by the load is approximately equal to the DC collector biasing resistor.

From a practical standpoint, input and output impedance cannot be measured directly with an ohmmeter. This is because ohmmeters measure resistance by sending out a small "sensing" current. The DC bias and AC signal currents will interact with this current and produce an unreliable result. Instead, impedances can be measured indirectly through a voltage divider effect. That is, if the voltages of both legs of a voltage divider can be measured and the resistance of one of the legs is known, the remaining resistance may be determined using Ohm's law or the voltage divider rule.

19.3 Experimental Example
19.3.1 Equipment
(1) Dual adjustable DC power supply model:_____ S/No.:_____
(1) DMM model:_____ S/No.:_____
(1) Dual channel oscilloscope model:_____ S/No.:_____
(1) Function generator model:_____ S/No.:_____
(3) Small signal transistors (2N3904)
(1) 10 kΩ resistor ¼ watt actual: _____
(1) 15 kΩ resistor ¼ watt actual: _____
(1) 20 kΩ resistor ¼ watt actual: _____
(1) 22 kΩ resistor ¼ watt actual: _____

(1) 33 kΩ resistor ¼ watt actual: _____
(2) 10 µF capacitors actual: _____
(1) 470 µF capacitor actual: _____

Schematic Diagrams

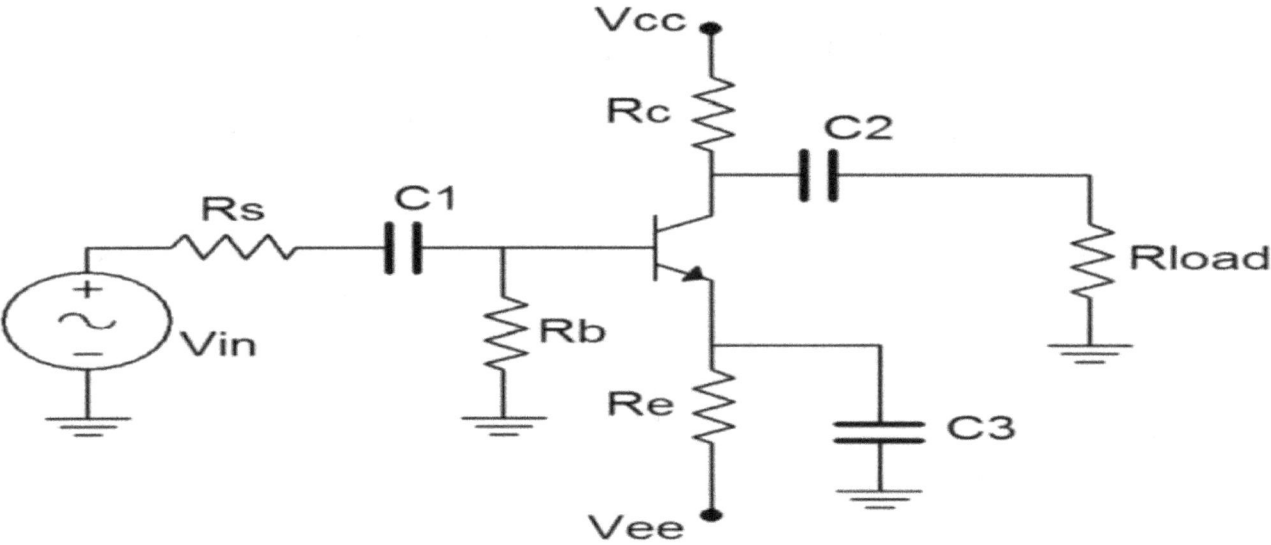

Figure 19.1

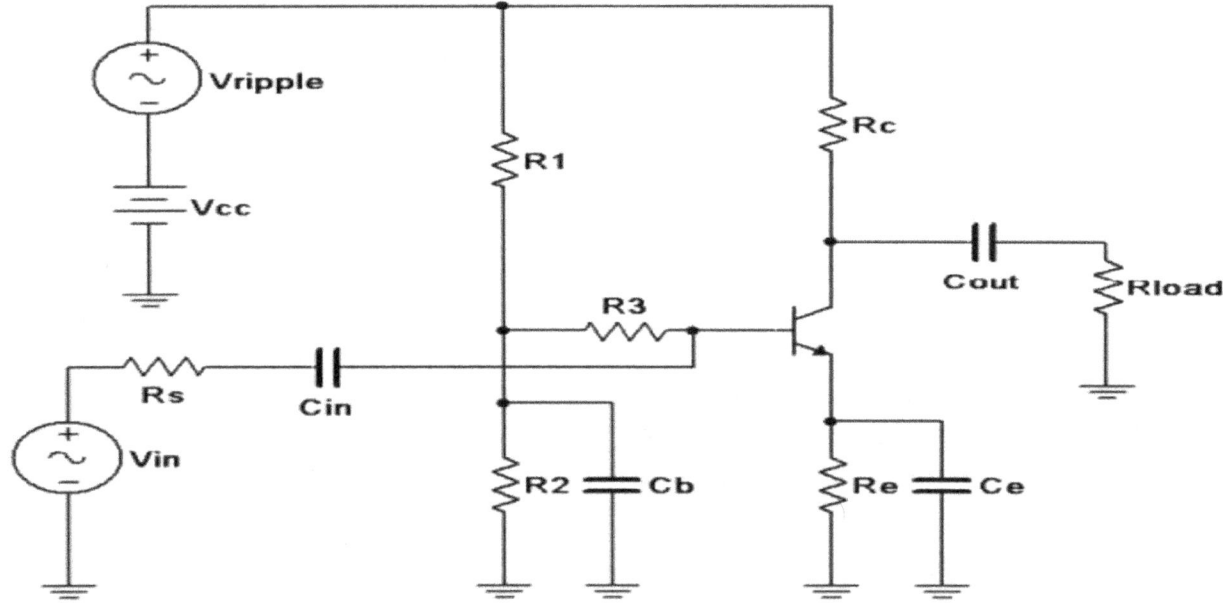

Figure 19.2

19.3.2 Procedure
DC Circuit Voltages

1. Consider the circuit of Figure 19.1 using Vcc = 15 volts, Vee = −12 volts, Rs = 10 kΩ, Rb = 33 kΩ, Re = 22 kΩ, Rc = 15 kΩ, R_{load} = 20 kΩ, C1 = C2 = 10 µF and C3 = 470 µF. Using the approximation of a negligible base voltage, determine the DC voltages at the base, emitter, and collector along with the collector current, and record these in Table 19.1.
2. Build the circuit of Figure 19.1 using Vcc = 15 volts, Vee = −12 volts, Rs = 10 kΩ, Rb = 33

kΩ, Re = 22 kΩ, Rc = 15 kΩ, R_{load} = 20 kΩ, C1 = C2 = 10 μF and C3 = 470 μF. Make sure that the AC source is turned off or disconnected. Measure the DC voltages at the base, emitter, and collector along with the collector current, and record these in Table 19.1. Note, you may wish to use a transistor curve tracer or beta checker to get approximate values of beta for each of the three transistors to be used.

AC Circuit Voltages

3. Based on the calculated collector current, determine the resulting theoretical r'_e, Av, Zin and Zout, and record these in Table 19.2. Assume a beta of approximately 150 for the Zin calculation.
4. Continuing with the values in Table 19.2 and using an AC source voltage of a 40 mV peak-peak 1 kHz sine wave, compute the theoretical AC base, emitter and load voltages, and record them in Table 19.3 (Theory). Note that Rs will create a voltage divider effect with Zin, thus reducing the signal that reaches the base. This reduced signal is then multiplied by the voltage gain and appears at the collector.
5. Set the source to a 40 mV peak-peak 1 kHz sine wave and apply to the circuit. Using the oscilloscope, place one probe at the base and the second at the emitter. Record the resulting peak-peak voltages in the first row of Table 19.3 (Experimental). The oscilloscope inputs should be set for AC coupling with the bandwidth limit engaged. Capture an image of the oscilloscope display.
6. Move the second probe to the load and record its peak-peak value in the first row of Table 19.3. Also include whether the signal is in phase or out of phase with the base signal. Capture an image of the oscilloscope display.
7. Unhook the load resistor from the output capacitor and measure the resulting collector voltage (do not connect the output capacitor to ground-simply leave it dangling). Record this value in the final column of Table 19.3.
8. Reattach the load resistor. Swap the transistor with the second transistor and repeat steps 5 through 7 using the second row of Table 19.3.
9. Reattach the load resistor. Swap the transistor with the third transistor and repeat steps 5 through 7 using the third row of Table 19.3.
10. Using the measured base and collector voltages from Table 19.3, determine the experimental gain for each transistor. From these gains determine the experimental r'_e. Using the source voltage, the measured base voltages and the source resistance, determine the effective input impedances via Ohm's law or the voltage divider rule. Finally, in similar manner and using the loaded and unloaded collector voltages along with the load resistor value, determine the experimental output impedances. Record these values in Table 19.4. Also determine and record the percent deviations.

Troubleshooting

11. Return the load resistor to the circuit. Consider each of the individual faults listed in Table 19.5 and estimate the resulting AC load voltage. Introduce each of the individual faults in turn and measure and record the load voltage in Table 19.5.

Computer Simulation

12. One issue with amplifiers is noise and ripple on the power supply. This will be directly coupled to output of the circuit via the collector resistor. Worse, this noise or ripple may be coupled into the base and then amplified along with the desired input signal. This can be an issue with amplifiers that use a voltage divider bias. One way to reduce

this effect is to decouple the voltage divider from the base. This modification is shown in the circuit of Figure 19.2. C_b effectively shorts R2, sending power supply noise and ripple to ground instead of into the base. By itself this would also short the desired input signal so an extra resistor, R3 is added between the capacitor and the base. The input impedance of the circuit is approximately equal to R3 in parallel with $\beta\, r'_e$. To show the effectiveness of this technique, build the circuit of Figure 2 in a simulator. Use values of Vin = 20 mV peak at 1 kHz, V_{ripple} = 20 mV peak at 120 Hz, Vcc = 12 volts, Rs = 1 kΩ, R1 = 10 kΩ, R2 = 3.3 kΩ, R3 = 22 kΩ, Re = 4.7 kΩ, Rc = 3.3 kΩ, R_{load} = 1 kΩ, C_{in} = C_{out} = 10 µF, Cb= 100 µF and Ce = 470 10 µF. Run a Transient simulation and look at the load voltage. A very small low frequency variation should be noted. This is the 120 Hz ripple coupled in through the collector resistor. Alter the circuit by removing C_b and R3 to produce the basic voltage divider circuit (or more simply, set C_b and R3 to extremely small values such as pF and mΩ). Rerun the simulation. The load voltage should now show a much more obvious ripple contribution, thus showing how effective the power supply decoupling components can be.

19.3.3 Result Tables

Table 19.1

$V_{B\,Theory}$	$V_{E\,Theory}$	$V_{C\,Theory}$	$I_{C\,Theory}$	$V_{B\,Exp}$	$V_{E\,Exp}$	$V_{C\,Exp}$	$I_{C\,Exp}$

Table 19.2

V'_C	A_v	Z_{in}	Z_{out}

Table 19.3

Transistor	$V_{B\,Theory}$	$V_{E\,Theory}$	$V_{L\,Theory}$	$V_{B\,Exp}$	$V_{E\,Exp}$	$V_{L\,Exp}$	Phase V_L	$V_{L\,No\,Load}$
1								
2								
3								

Table 19.4

Transistor	$A_{v\,Exp}$	r'_e	$Z_{in\,Exp}$	$Z_{out\,Exp}$	%D A_v	%D r'_e	%D Z_{in}	%D Z_{out}
1								
2								

3								

Table 19.5

Issue	V_{Load}
R_B Short	
C_1 Open	
R_C Short	
R_C Open	
R_E Open	
C_2 Open	
C_3 Open	
V_{CE} Open	

Questions

1. Does the common emitter amplifier produce a considerable amplification effect and if so, are the results consistent across transistors?

   _____
   _____
   _____
   _____
   _____

2. Does the common emitter amplifier produce a phase shift at the output and if so, is it affected by the transistor beta?

   _____
   _____
   _____
   _____
   _____

3. If the collector and base voltages had been measured with the oscilloscope DC coupled, how would the measurements of Table 19.3 have changed?

   _____
   _____

4. Does the value of the transistor beta play any role in setting the input impedance? Was a considerable variation in input impedance apparent?

CHAPTER TWENTY
Swamped CE Amplifier

20.1 Objective
The objective of this exercise is to examine the characteristics of a swamped common emitter amplifier, specifically the effects of swamping on voltage gain, input impedance and distortion.

20.2 Theory Overview
As the signal current changes in a transistor, the total current flowing through the emitter changes along with it. As a result, these changes produce small changes in the internal emitter resistance r'_e which in turn changes the voltage gain. In other words, the gain changes throughout the signal producing slightly more or less gain at some points along the signal than others. These changes show up as a squashing or elongating of the positive and negative peaks of the output signal. Generally, these forms of waveform distortion are to be avoided.

Also, they tend to worsen as the output signal amplitude increases. A method of mitigating this distortion is to add AC resistance to the emitter portion of the circuit. This added resistance tends to buffer or "swamp out" the changes in r'e and therefore reduces the distortion. A side bonus is that Zin(base) will also be increased which will result in an increased Zin to the circuit. On the downside, the added resistance will lower the voltage gain. Consequently the swamped amplifier exhibits a lower gain but one of higher quality. In general, the larger the swamping resistance is compared to r'e, the greater the effects on distortion, gain and input impedance.

20.3 Experimental Example
20.3.1 Equipment
 (1) Dual adjustable DC power supply model:_____ S/No.:_____
 (1) DMM model:_____ S/No.:_____
 (1) Dual channel oscilloscope model:_____ S/No.:_____
 (1) Low distortion function generator model:_____ S/No.:_____
 (1) Distortion analyzer model:_____ S/No.:_____
 (1) Small signal transistor (2N3904)
 (1) 220 Ω resistor ¼ watt actual: _____
 (1) 1 k Ω resistor ¼ watt actual: _____
 (1) 10 k Ω resistor ¼ watt actual: _____
 (1) 15 k Ω resistor ¼ watt actual: _____
 (1) 20 k Ω resistor ¼ watt actual: _____
 (1) 22 k Ω resistor ¼ watt actual: _____
 (1) 33 k Ω resistor ¼ watt actual: _____
 (2) 10 µF capacitors actual: _____
 (1) 470 µF capacitor actual: _____

Schematic Diagrams

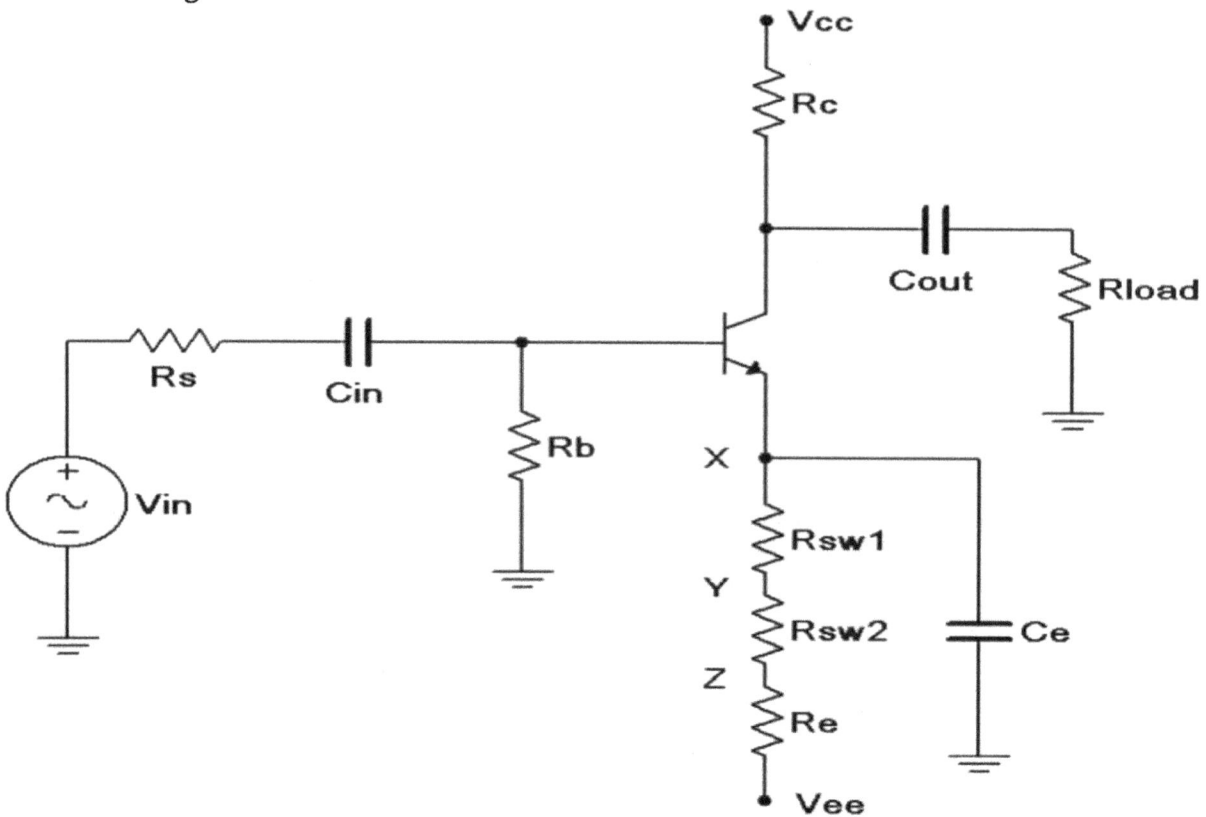

Figure 20.1

20.3.2 Procedure
AC Circuit Voltages

1. Consider the circuit of Figure 20.1 using Vcc = 15 volts, Vee = -12 volts, Rs = 10 kΩ, Rb = 33 kΩ, Re = 22 kΩ, R_{sw1} = 220 Ω, R_{sw2} = 1 kΩ, Rc = 15 kΩ, Rload = 20 kΩ, Cin = Cout = 10 µF and C_e=470 µF. Using the approximation of a negligible DC base voltage, determine the DC collector current and r'_e, and record these in Table 20.1. Using the r'_e, calculate the expected Z_{in}, Z_{in}(base), and Av for the X, Y and Z connection points for Ce (shown at position X in the schematic). Record these in Table 20.2. If a transistor curve tracer or beta checker is not available to get an approximate value of beta for the transistor, estimate it at 150.

2. Build the circuit of Figure 1 using Vcc = 15 volts, Vee = -12 volts, Rs = 10 kΩ, Rb = 33 kΩ, Re=22 kΩ, R_{sw1} = 220 Ω, R_{sw2} = 1 kΩ, Rc = 15 kΩ, R_{load} = 20 kΩ, C_{in} = C_{out} = 10 µF and Ce=470µF. Connect C_e to position X. Disconnect the signal source and check the DC transistor voltages to ensure that the circuit is biased correctly (note, the DC equivalent circuit is very similar to the ones used in the Emitter Bias Exercise and should exhibit similar DC voltage readings).

3. Using a 1 kHz sine wave setting, apply the signal source to the amplifier and adjust it to achieve a load voltage of 2 volts peak-peak.

4. Measure the AC peak-peak voltages at the source, the base, and the load, and record these in Table 20.3. The load waveforms may exhibit some asymmetry due to distortion so be sure to record the peak-peak voltage not the peak. If asymmetry is

observed between the positive and negative peaks, make a note of it. Also, capture images of the oscilloscope displays (Vs with Vb and Vb with V_{load}).

5. Set the distortion analyzer to 1 kHz and % total harmonic distortion (% THD). Apply it across the load and record the resulting reading in the final column of Table 20.3.
6. Remove the distortion analyzer and connect Ce to position Y instead of X. Repeat steps 3, 4 and 5.
7. Remove the distortion analyzer and connect Ce to point Z instead of Y. Repeat steps 3, 4 and 5.
8. Using the measured base and load voltages from Table 20.3, determine the experimental gain for the transistor. Using the measured source and base voltages along with the source resistance, determine the effective input impedances via Ohm's law or the voltage divider rule. Record these values in Table 20.4. Also determine and record the percent deviations.

Computer Simulation

9. Build the circuit in a simulator and run three sets of simulations, one for each of the three Ce positions. For each trial, set the AC source voltage to the value measured in Table 20.3 (VS Exp). Run a Transient Analysis and inspect the voltages at the base and load. The AC source voltage may have to be adjusted slightly to achieve the desired 2 volt peak-peak load voltage. Record these values in Table 20.5. Add the Distortion Analyzer instrument at the load and record the resulting value.

20.3.3 Result Tables

Table 20.1

	I_C	
	r'_e	

Table 20.2

Position	$A_{v\,Teory}$	$Z_{in(Base)\,Teory}$	$Z_{in\,Teory}$
X			
Y			
Z			

Table 20.3

Position	$V_{s\,Exp}$	$V_{B\,Exp}$	$V_{L\,Exp}$	%THD
X				
Y				
Z				

Table 20.4

Position	$A_{v\,Exp}$	$Z_{in\,Exp}$	% Dev A_v	% Dev Z_{in}

	X				
	Y				
	Z				

Table 20.5

Position	$V_{S\,Sim}$	$V_{B\,Sim}$	$V_{L\,Sim}$	% Distorsion
X				
Y				
Z				

20.3.4 Questions

1. In summary, what are the effects of swamping?

2. Is the change in voltage gain directly proportional to the amount of swamping?

3. Is the change in input impedance directly proportional to the amount of swamping?

4. Is the change in distortion directly proportional to the amount of swamping?

5. Are THD levels below 1% easily discerned on a simple oscilloscope display?

6. Why is it important that the load voltage be set to the same value in each of the three trials instead of setting the source to the same value?

CHAPTER TWENTY-ONE
Frequency Limits

21.1 Objective
This exercise focuses on the analysis of the frequency limits of a transistor amplifier. The elements contributing to both the lower and upper limits of frequency performance are examined, thus defining the mid-band region of the circuit.

21.2 Theory Overview
As the signal frequency extends to very high or very low frequencies, capacitive effects on gain can no longer be ignored or idealized as shorts or opens. At low frequencies, coupling and bypass capacitors in conjunction with surrounding resistance create lead networks that cause a reduction in voltage gain. At higher frequencies, small shunting capacitances associated with individual devices circuit wiring create lag networks. These will also create a reduction in voltage gain. In general, the highest critical frequency among the lead networks creates the amplifier's lower limit frequency, f1. In contrast, the lowest critical frequency among the lag networks creates the amplifier's upper limit frequency, f2. These points are defined as the half-power points and can be determined experimentally by finding those frequencies at which the output voltage (and hence, voltage gain) has fallen to 70.7% of the mid-band value. The values are found theoretically by Thevenizing the circuitry around the capacitor in question, reducing it to a single resistance, and solving for the critical frequency, f_c.

21.3 Experimental Example
21.3.1 Equipment
(1) Dual adjustable DC power supply model:_____ S/No.:_____
(1) DMM model:_____ S/No.:_____
(1) Dual channel oscilloscope model:_____ S/No.:_____
(1) Function generator model:_____ S/No.:_____
(1) Small signal transistor (2N3904)
(1) 10 k Ω resistor ¼ watt actual: _____
(1) 15 k Ω resistor ¼ watt actual: _____
(1) 20 k Ω resistor ¼ watt actual: _____
(1) 22 k Ω resistor ¼ watt actual: _____
(1) 33 k Ω resistor ¼ watt actual: _____
(1) 2.2 nF capacitor actual: _____
(1) 10 nF capacitor actual: _____
(2) 10 µF capacitors actual: _____
(1) 470 µF capacitor actual: _____

Schematic Diagram

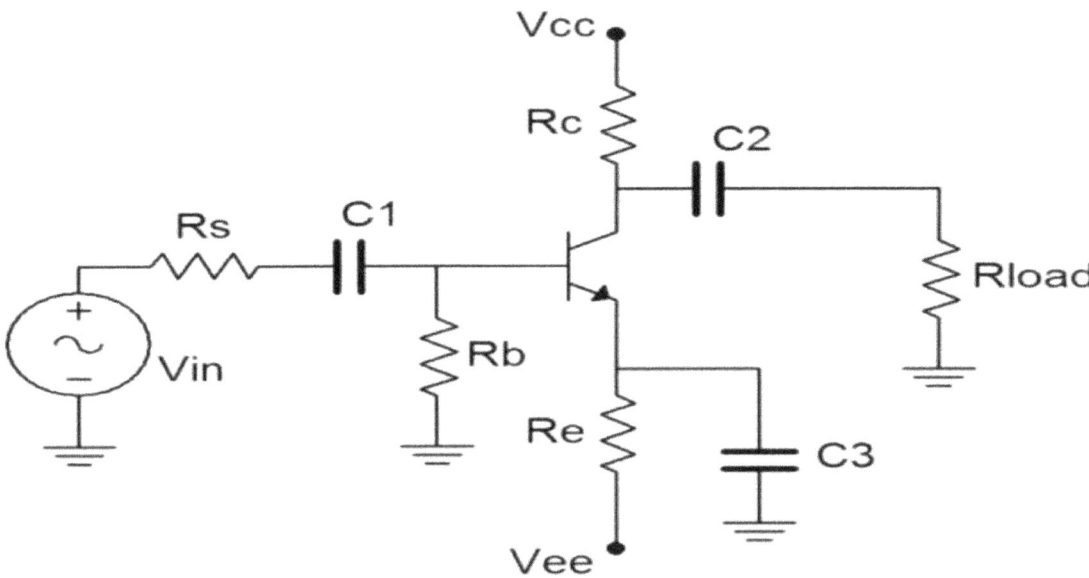

Figure 21.1

21.3.2 Procedure

Mid-band Response

1. The circuit of Figure 21.1 is the same as the one used in the Common Emitter Amplifier exercise. The values used were Vcc = 15 volts, Vee = -12 volts, Rs = 10 kΩ, Rb = 33 kΩ, Re = 22 kΩ, Rc = 15 kΩ, R_{load} = 20 kΩ, C1 = C2 = 10 µF and C3 = 470 µF. It was shown that the amplifier produced considerable voltage at 1 kHz. Build the circuit using these values and verify that it is operating correctly by setting Vin to a 40 mV peak-peak sine wave at 1 kHz and measuring Vout. Compute the voltage gain and record these two values in Table 21.1.
2. Compute the expected critical frequencies of the input and output coupling networks along with the emitter bypass network and record the values in Table 21.2. Include the Thevenized resistance for each equivalent circuit.

Lower Frequency Limit

3. The input coupling network can be made clearly dominant by replacing C1 with a smaller value. Decreasing C1 from 10 µF to 10 nF will increase its critical frequency by a factor of 1000. Replace C1 with this value.
4. Set the AC source voltage to a 40 mV peak-peak 10 kHz sine wave. Note the output level at the load. Sweep the frequency between 5 kHz and 20 kHz to verify that the gain is stable. Decrease the input frequency until the load signal drops to 70.7% of the 10 kHz level. Record this value in Table 3.
5. The output network may be examined in a similar fashion. Return C1 to 10 µF, replace C2 with a
10 nF and repeat step 4.
6. Return C2 to the original 10 µF capacitor before proceeding.

Upper Frequency Limit

7. The upper frequency limit is controlled by small device and wiring capacitances that will vary with the precise components used and the circuit layout. To minimize potential errors, a large load capacitance can be shunted across R_{load} to bring the critical frequency down to an easily managed frequency. This could also represent the

effect of cable capacitance.
8. Place a 2.2 nF capacitor across the load. Compute the effective resistance of the lag network and its corresponding critical frequency in Table 21.4.
9. Set the AC source voltage to a 40 mV peak-peak 1 kHz sine wave. Note the output level at the load. Sweep the frequency around 1 kHz to verify that the gain is stable. Increase the input frequency until the load signal drops to 70.7% of the 1 kHz level. Record this value in Table 21.4.

Computer Simulation
10. Perform an AC Analysis (Bode plot) of the amplifier using the original capacitor values and for the three variations. Plot the gain from 1 Hz to 10 MHz for the original circuit and from 10 Hz to 100 kHz for the three variations. Compare the simulated results to the experimental values.

21.3.3 Result Tables
Table 21.1

V_{in}	V_{out}	A_v

Table 21.2

Network	$R_{Thevenin}$	f_C
Input		
Output		
Bypass		

Table 21.3

Capacitor	$f_{C\,Theory}$	$f_{C\,Exp}$	% Dev
C1 = 10 nF			
C2 = 10 nF			

Table 21.4

R_{out}	$f_{out\,Theory}$	$f_{out\,Exp}$	% Dev

21.3.4 Questions
1. What effect might resistor and capacitor tolerance have on critical frequencies?

_____
_____
_____
_____

2. Does beta variation have an impact on frequency response? If so, how?

_____

3. Explain the possible effects of load impedance on the frequency response of the amplifier.

CHAPTER TWENTY-TWO
Voltage Follower

22.1 Objective

The objective of this exercise is to examine the characteristics of a voltage follower, specifically an emitter follower using a Darlington pair. Voltage gain, input impedance and distortion will all be examined.

22.2 Theory Overview

The function of a voltage follower is to present high input impedance and low output impedance with a non-inverting gain of one. This allows the load voltage to accurately track or follow the source voltage in spite of a large source/load impedance mismatch. Ordinarily this mismatch would result in a large voltage divider loss. Consequently, followers are often used to drive a low impedance load or to match a high impedance source. While typical laboratory sources exhibit low internal impedances, some circuits and passive transducers can exhibit quite high internal impedances. For example, electric guitar pickups can exhibit in excess of 10 k Ω at certain frequencies. Although the voltage gain may be approximately one, current gain and power gain can be quite high, especially if a Darlington pair is used. Besides unity voltage gain and a high Z_{in} and low Z_{out}, followers also tend to exhibit low levels of distortion.

The Darlington pair effectively produces a "beta times beta" effect by feeding the emitter current of one device into the base of a second transistor. This also produces the effect of doubling both the effective V_{BE} and r'_e.

22.3 Experimental Example

22.3.1 Equipment

(1) Dual adjustable DC power supply model:_____ S/No.:_____
(1) DMM model:_____ S/No.:_____
(1) Dual channel oscilloscope model:_____ S/No.:_____
(1) Low distortion function generator model:_____ S/No.:_____
(1) Distortion analyzer model:_____ S/No.:_____
(2) Small signal transistors (2N3904)
(1) 220 Ω resistor ¼ watt actual: _____
(1) 1 k Ω resistor ¼ watt actual: _____
(1) 22 k Ω resistor ¼ watt actual: _____
(1) 470 k Ω resistor ¼ watt actual: _____
(1) 10 µF capacitor actual: _____
(1) 470 µF capacitor actual: _____

Schematic Diagram

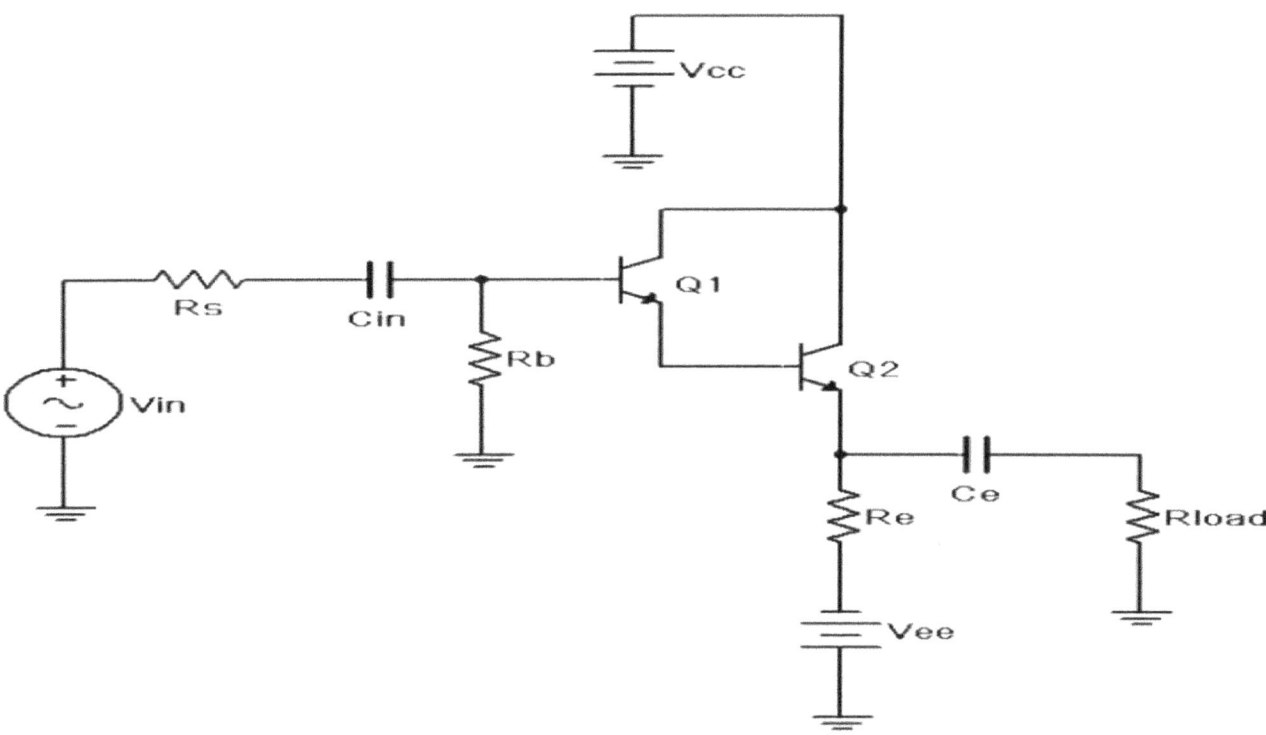

Figure 22.1

22.3.2 Procedure

1. Consider the circuit of Figure 22.1 using Vcc = 5 volts, Vee = -12 volts, Rs = 22 kΩ, Rb = 470 kΩ, Re = 1 kΩ, R_{load} = 220 Ω, Cin = 10 μF and Ce = 470 μF. Using the approximation of a negligible DC base voltage, determine the DC collector current and r'e, and record these in Table 1. Using the r'_e, calculate the expected Zin, Zin(base), Z_{out} and A_v. Record these in Table 22.2. If a transistor curve tracer or beta checker is not available to get an approximate value of beta for the transistors, estimate the pair at 10,000.
2. Build the circuit of Figure 1 using Vcc = 5 volts, Vee = -12 volts, Rs = 22 kΩ, Rb = 470 kΩ, Re=1kΩ, R_{load} = 220 Ω, Cin = 10 μF and Ce = 470 μF. Disconnect the signal source and check the DC transistor voltages to ensure that the circuit is biased correctly. (Note: The base should be close to zero while the emitter will be two V_{BE} drops less, or about -1.4 V_{DC}.)
3. Using a 1 kHz sine wave setting, apply the signal source to the amplifier and adjust it to achieve a source voltage of 2 volts peak-peak (i.e., to the left of Rs).
4. Measure the AC peak-peak voltages at the source, the base, and the load, and record these in Table 22.3. Also note the phase of the load voltage compared to the source. If distortion asymmetry is observed between the positive and negative peaks, make a note of it. Also, capture images of the oscilloscope displays (Vs with Vb and Vb with V_{load}).
5. Set the distortion analyzer to 1 kHz and % total harmonic distortion (% THD). Apply it across the load and record the resulting reading in Table 22.3.
6. Finally, unhook (i.e., open) the load and measure the resulting load voltage. Record this in the final column of Table 22.3.
7. Using the measured base and load voltages from Table 22.3, determine the experimental gain for the circuit. Using the measured source and base voltages along

with the source resistance, determine the effective input impedance via Ohm's law or the voltage divider rule. In similar fashion, using the loaded and unloaded load voltages along with the load resistance, determine the effective output impedance. Record these values in Table 22.4. Also determine and record the percent deviations.

Troubleshooting

8. Return the load resistor to the circuit. Consider each of the individual faults listed in Table 22.5 and estimate the resulting AC load voltage. Introduce each of the individual faults in turn and measure and record the load voltage in Table 22.5.

Computer Simulation

9. Build the circuit in a simulator and run a Transient Analysis. Use a 1 kHz 1 volt peak sine for the source. Inspect the voltages at the source, base and load. Record these values in Table 22.6. Add the Distortion Analyzer instrument at the load and record the resulting value.

21.3.3 Result Tables

Table 22.1

	I_C	
	r'_e	

Table 22.2

$A_{v\,Teory}$	$Z_{in(Base)\,Teory}$	$Z_{in\,Teory}$	$Z_{out\,Teory}$

Table 22.3

$V_{s\,Exp}$	$V_{B\,Exp}$	$V_{L\,Exp}$	Phase V_L	%THD	$V_{L\,No\,Load}$

Table 22.4

$A_{v\,Exp}$	$Z_{in\,Exp}$	$Z_{out\,Exp}$	% Dev A_v	% Dev Z_{in}	% Dev Z_{out}

Table 22.5

Issue	V_{Load}
R_b Short	
C_{in} Open	
R_e Open	
C_e Open	

R_{Load} Short	
V_{CE} Open	
V_{CC} Open	

Table 22.6

$V_{S\,Sim}$	$V_{B\,Sim}$	$V_{L\,Sim}$	% Distortion

22.3.4 Questions

1. Would the results of this exercise have been considerably different if the load had been ten times larger? What does that say about the performance of the circuit?

2. If the 22 k source had been directly connected to the 220 load without the follower in between, what would be the load voltage?

3. How do the THD levels of the follower compare to those of the swamped common emitter amplifier?

4. How would the circuit parameters change if a Darlington pair had not been used?

CHAPTER TWENTY-THREE
Class 'A' Power Analysis

23.1 Objective
The objective of this exercise is to examine large signal class A operation. A voltage follower will be investigated by plotting the AC load line and determining output compliance, maximum load power, supplied DC power and efficiency. The effects of clipping will be noted.

23.2 Theory Overview
The maximum output signal, or compliance, of a class A amplifier is determined by its AC load line. The maximum peak level is determined by the smaller of V_{CEQ} and $I_{CQ} \cdot r_{Load}$. If either of these levels is hit, the output signal will begin to clip causing greatly increased distortion. Knowing this voltage and the load resistance, the maximum load power may be determined. Dividing this power by the total supplied DC power will yield the efficiency. The maximum theoretical efficiency of an R_C coupled class A amplifier is 25% although real-world circuits may be far less. In fact, the power dissipation of the transistor itself (P_{DQ}) may be greater than the maximum load power, clearly not a desirable condition. Note that the total supplied power is the product of the total supplied voltage and the average total current. In a class A amplifier that is not clipping, the average supplied current is equal to the quiescent DC current. In the case of a dual supply emitter biased circuit, this is simply the collector current and can be measured with a DC ammeter.

23.3 Experimental Example
23.3.1 Equipment
(1) Dual adjustable DC power supply model:_____ S/No.:_____
(1) DMM model:_____ S/No.:_____
(1) Dual channel oscilloscope model:_____ S/No.:_____
(1) Low distortion function generator model:_____ S/No.:_____
(1) Distortion analyzer model:_____ S/No.:_____
(2) Small signal transistors (2N3904)
(1) 100 Ω resistor ¼ watt actual: _____
(1) 1 k Ω resistor ¼ watt actual: _____
(1) 47 k Ω resistor ¼ watt actual: _____
(1) 470 µF capacitor actual: _____
Schematic Diagram

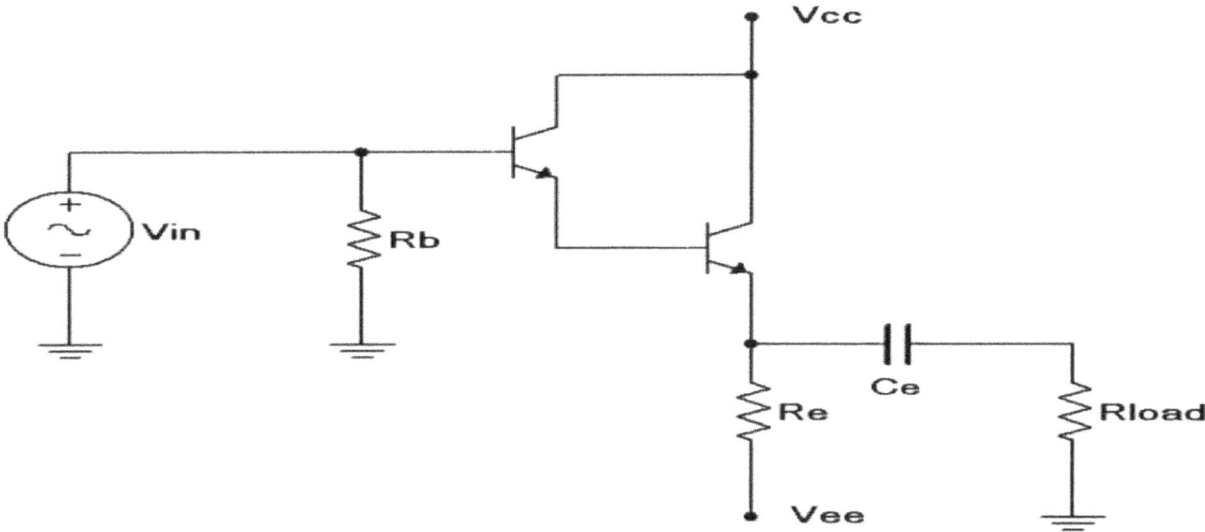

Figure 23.1

23.3.2 Procedure

AC Load Line and Power Analysis

1. Consider the circuit of Figure 23.1 using Vcc = 5 volts, Vee = -12 volts, Rb = 47 kΩ, Re = 1 kΩ, R_{load} = 100 Ω and Ce = 470 μF. Determine the theoretical I_{CQ}, V_{CEQ}, $V_{CE(cutoff)}$ and $i_{C(sat)}$, and record these in Table 23.1. It is helpful to plot the AC load line for step three. Note that the collector-emitter saturation voltage for a Darlington pair cannot be assumed to be 0 volts, and may be closer to one volt, thus reducing the expecting voltage swing toward the saturation point. It is also worth noting that this amplifier has a direct coupled input (i.e., no input capacitor is required due to the very small DC base voltage).

2. Build the circuit of Figure 23.1 using Vcc = 5 volts, Vee = -12 volts, Rb = 47 kΩ, Re = 1 kΩ, R_{load}=100 Ω and Ce = 470 μF. Disconnect the signal source and measure the DC transistor voltages to ensure the circuit is biased correctly. Record V_{CEQ} and I_{CQ} in Table 23.1 (Experimental).

3. Based on the data recorded in Table 23.1, determine the theoretical maximum unclipped load voltage (compliance) and record it in Table 23.2. Based on this, determine the maximum load power and record it Table 23.2 as well. Also determine and record the expected values for the quiescent power dissipation of the transistor (P_{DQ}), the supplied DC current and power, and the resulting efficiency.

4. Using a 1 kHz sine wave setting, apply the signal source to the amplifier and adjust it to achieve a load voltage that just begins to clip. Reduce the amplitude slightly to produce a clean, unclipped wave. Record this level as the experimental compliance in Table 23.2. From this, determine and record the experimental maximum load power. Also, capture an image of the oscilloscope display.

5. Insert an ammeter in the collector and measure the resulting current with the signal still set for maximum unclipped output. Record this in Table 23.2 as $I_{supplied}$ (Experimental).

6. Using the data already recorded, determine and record the experimental P_{DQ}, $P_{Supplied}$, and η. Finally, determine the deviations for Table 23.2.

Clipping and Distortion

7. Increase the signal until both peaks begin to clip. Record these clipping levels in Table 23.3. Make sure the oscilloscope is DC coupled for this measurement as any offset is important. Compare these peaks to those predicted by the AC load line. Also, capture an image of the oscilloscope display.
8. Decrease the signal level so that it is about 90% of the maximum unclipped level. Set the distortion analyzer to 1 kHz and % total harmonic distortion (% THD). Apply it across the load and record the resulting reading in Table 23.4 (Normal). Increase the signal by about 20% so that one of the peaks is obviously clipped and take a second distortion reading, recording it Table 23.4 (Clipped).

Computer Simulation

9. Build the circuit in a simulator and run a Transient Analysis. Use a 1 kHz 7 volt peak sine for the source. Inspect the voltage at the load. Record the peak clip points in Table 23.5. Reduce the input signal so that clipping disappears. If available, add the Distortion Analyzer instrument at the load and record the resulting value.

23.3.3 Result Tables

Table 23.1

	Theory	Experimental
I_{CQ}		
V_{CEQ}		
$i_{C(Sat)}$		X
$V_{CE(cutoff)}$		X

Table 23.2

	Theory	*Experimental*	*% Deviation*
Compliance			
$P_{Load(max)}$			
$I_{Supplied}$			
P_{DQ}			
$P_{Supplied}$			
η			

Table 24.4

% THD Normal	
% THD Clipped	

Table 23.3

Positive Clip	
Negative Clip	

Table 23.5

Positive Clip	Negative Clip	% Deviation

13.3.4 Questions

1. Does the maximum load power compare favorably to the supplied DC power and the transistor's power dissipation? That is, is the circuit efficient?

2. How does the THD level of the clipped signal compare to that of the unclipped signal?

3. How well do the clip levels measured and simulated compare to the predicted AC load line?

4. How would the circuit performance change if a Darlington pair had not been used? Would this affect the AC load line?

5. Would increasing the Vcc supply increase the output compliance? Why/why not?

CHAPTER TWENTY-FOUR
Class 'B' Power Analysis

24.1 Objective
The objective of this exercise is to examine large signal class B operation. A voltage follower will be investigated to determine output compliance, maximum load power, supplied DC power and efficiency.
The effects of crossover distortion will be noted by comparing resistor and diode biasing schemes.

24.2 Theory Overview
The maximum output signal, or compliance, of a class B amplifier is determined by its AC load line. The peak to peak compliance is roughly equal to the total DC supply voltage(s). As two output devices are used, each conducting for half of the cycle, the quiescent current can remain low, unlike a class A amplifier. This results in vastly improved efficiency, theoretically up to 78.5%. The switchover from one transistor to the other is problematic and can result in crossover or notch distortion. To alleviate this, the transistors are given a small idle current so that each base-emitter junction is just about fully on. While resistors can be used to create this bias, trying to match the linear current-voltage characteristic of a resistor to the logarithmic characteristic of a PN junction is tricky. Consequently, another PN junction, namely a diode, is used instead. The diode will result in a more stable circuit which produces less notch distortion.

24.3 Experimental Example
24.3.1 Equipment
(1) Dual adjustable DC power supply model:_____ S/No.:_____
(1) DMM model:_____ S/No.:_____
(1) Dual channel oscilloscope model:_____ S/No.:_____
(1) Low distortion function generator model:_____ S/No.:_____
(1) Distortion analyzer model:_____ S/No.:_____
(1) Small signal NPN transistor (2N3904)
(1) Small signal PNP transistor (2N3906)
(2) Switching diodes (1N914 or 1N4148)
(1) 100 Ω resistor ¼ watt actual: _____
(2) 220 Ω resistors ¼ watt actual: _____
(2) 2.2 kΩ resistors ¼ watt actual: _____
(2) 10 µF capacitor actual: _____
(1) 100 µF capacitor actual: _____

154 Laboratory Manual for Semiconductor Devices: Theory and Application

Schematic Diagrams

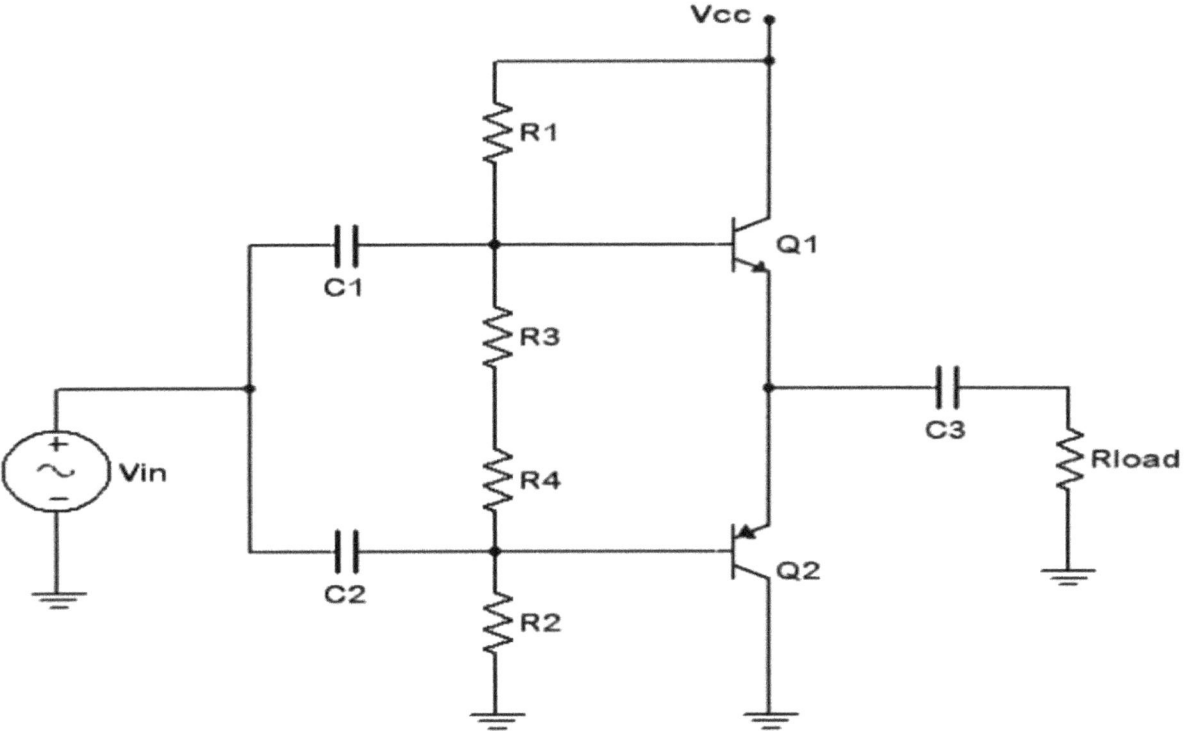

Figure 24.1

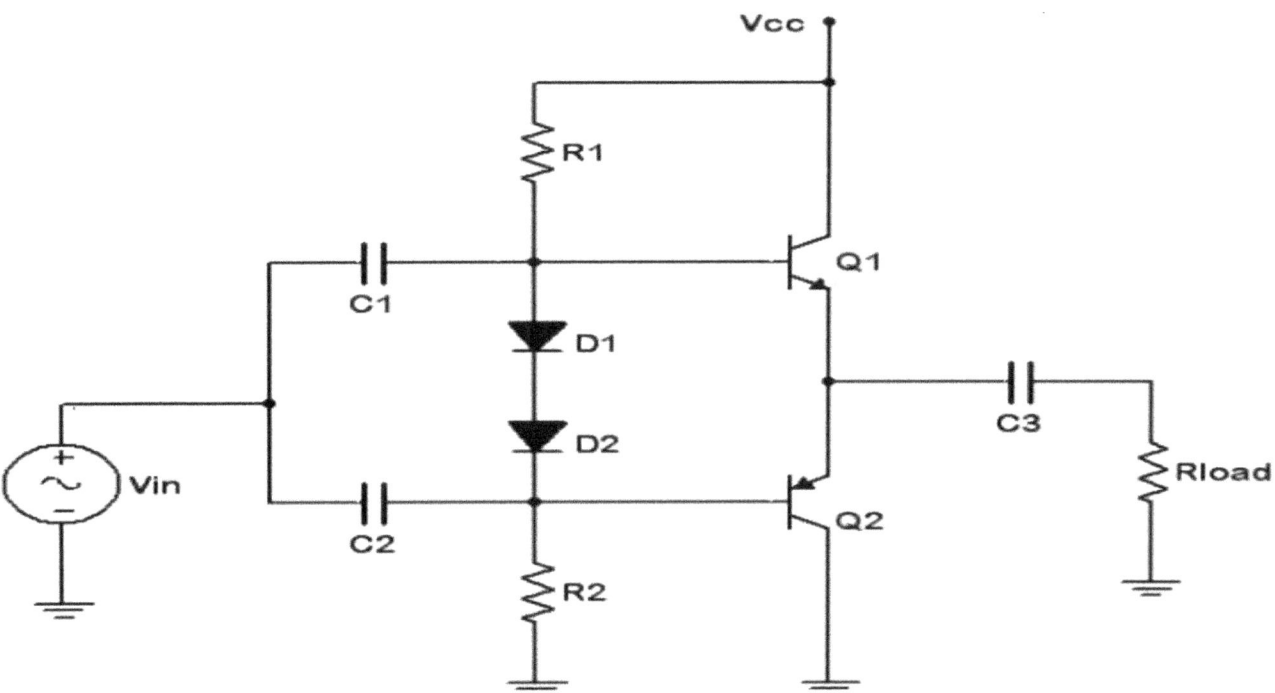

Figure 24.2

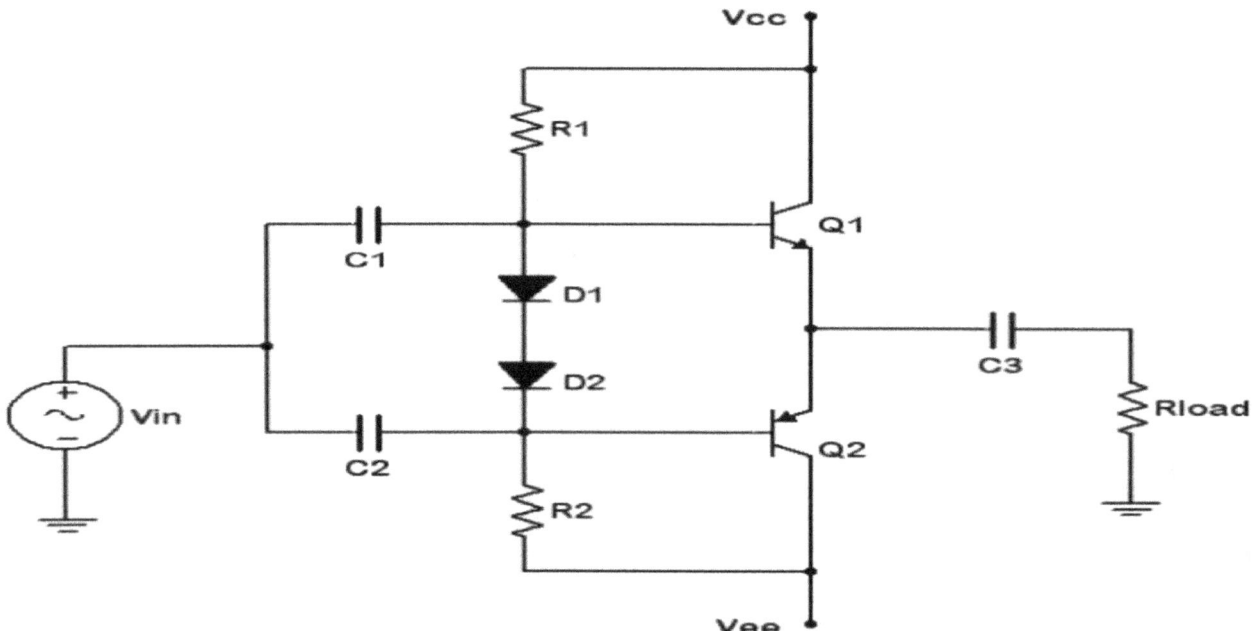

Figure 24.3

24.3.2 Procedure
Resistor versus Diode Bias and Crossover Distortion
1. Consider the circuit of Figure 24.1 using Vcc = 6 volts, R1 = R2 = 2.2 kΩ, R3 = R4 = 220 Ω, R_{load} = 100 Ω, C1 = C2 = 10 µF and C3 = 100 µF. Ideally, this circuit will produce a compliance of just under 6 volts peak-peak.
2. Build the circuit of Figure 1 using Vcc = 6 volts, R1 = R2 = 2.2 kΩ, R3 = R4 = 220 Ω, R_{load} = 100Ω, C1 = C2 = 10 µF and C3 = 100 µF. Disconnect the signal source and insert an ammeter into the collector of Q1. Record ICQ in Table 24.1.
3. Connect the signal source and apply a 1 kHz sine at 2 volts peak. Look at the load voltage and capture the oscilloscope image. There should be considerable notch or crossover distortion.
4. Cycle through the remaining supply voltages in Table 24.1, repeating steps 2 and 3. Only images of the first and last trials need be captured. As the bias current increases, the notch distortion should decrease.
5. Replace R3 and R4 with switching diodes, as shown in Figure 24.2. Repeat steps 2 through 4 using this circuit and Table 24.2. Overall, the superior matching of the diodes to the transistors should result in decreased notch distortion.

Dual Supply and Power Analysis
6. Add the negative power supply so that the circuit now appears as Figure 24.3. Set the power supplies to +/-6 volts DC. This should produce similar bias and amplification results to the single 12 volt supply circuit of Figure 24.2. Although the output coupling capacitor is no longer needed (one advantage of the dual supply topology), leave it in for safety sake.
7. Based on the I_{CQ} recorded for the 12 volt supply in Table 24.2, determine the theoretical P_{DQ}. Also determine the expected compliance, $P_{Load(max)}$, $I_{supplied}$, $P_{supplied}$ and efficiency. Record these values in the Theoretical column of Table 24.3.
8. Apply the signal source to the amplifier and adjust it to achieve a load voltage that

just begins to clip. Reduce the amplitude slightly to produce a clean, unclipped wave. Record this level as the experimental compliance in Table 24.3. From this, determine and record the experimental maximum load power. Also, capture an image of the oscilloscope display.

9. Insert an ammeter in the collector and measure the resulting current with the signal still set for maximum unclipped output. Record this in Table 24.3 as $I_{supplied}$ (Experimental). Remove the ammeter.

10. Using the data already recorded, determine and record the experimental P_{DQ}, $P_{Supplied}$, and η. Finally, determine the deviations for Table 24.3.

Distortion

10. Unlike class 'A' distortion which gets worse as the signal increases, notch distortion is relatively fixed. Therefore, it represents a smaller percentage of the overall output signal as the signal increases. To see this effect, adjust the signal level to achieve a load voltage of 8 volts peak-peak. There should be no clipping. Set the distortion analyzer to 1 kHz and % Total Harmonic Distortion (% THD). Apply it across the load and record the resulting reading in Table 24.4 (8 Vpp). Decrease the generator to achieve a load voltage of 1 volt peak-peak and record the resulting THD.

Computer Simulation

11. Build the circuit in a simulator and run a Transient Analysis. Use a 1 kHz 7 volt peak sine for the source. Inspect the voltage at the load. Record the peak clip points in Table 24.5. Reduce the input signal so that clipping disappears. Add the Distortion Analyzer instrument at the load and record the resulting value.

Table 24.2

Supply	I_{CQ} – Diodes
6 V	
8 V	
10 V	
12 V	

24.3.3 Result Tables

Table 24.1

Supply	I_{CQ} – Resistors
6 V	
8 V	
10 V	
12 V	

Table 24.3

	Theory	Experimental	% Deviation
Compliance			
$P_{Load(max)}$			
$I_{Supplied}$			
P_{DQ}			
$P_{Supplied}$			
η			

Table 24.5

Positive Clip	Negative Clip	% Distortion

Table 24.4

% THD 8 Vpp	I_{CQ} – Resistors
% THD 1 Vpp	

24.3.4 Questions

1. Does the maximum load power compare favorably to the supplied DC power and the transistor's power dissipation? That is, is the circuit efficient? How does it compare to class A operation?

   _____
   _____
   _____
   _____

2. How is the notch distortion affected by the power supply?

   _____
   _____
   _____
   _____

3. Compare the resistor bias and diode bias circuits regarding idle current (ICQ) and notch distortion. Compute the I_{CQ} versus V_{CC} stability (I_{CQ-MAX} / I_{CQ-MIN}) of each circuit using the first and last entries of Tables 24.1 and 24.2.

   _____
   _____
   _____
   _____

4. How does the class B circuit distortion compare to class 'A' operation?

5. Would increasing the Vcc supply increase the output compliance? Why/why not?

CHAPTER TWENTY-FIVE
Power Amp with Driver

25.1 Objective
The objective of this exercise is to examine a typical audio amplifier consisting of a class 'A' driver feeding a class B follower. System gain and clipping limits will be examined along with the audibility of clipping distortion and the shapes of voice waveforms.

25.2 Theory Overview
Typical audio amplifiers utilize one or smaller signal class 'A' stages to achieve sufficient voltage gain which then feeds a class B power stage connected to the load (normally a loudspeaker). The stage preceding the power section is referred to as the driver stage or simply the driver. The driver is often directly coupled instead of coupled via a capacitor. This maximizes gain and reduces component count.

A typical loudspeaker exhibits nominal 8 Ω impedance. As such, it demands considerable current. The job of the class B follower is to create a good match to this low impedance and produce sufficient current and power gain to drive it effectively. The voltage gain comes from the prior stages. If any of the amplifier stages clip the waveform, the loudspeaker will reproduce the distorted wave. This distortion can be clearly audible and produce a signal that sounds fuzzy or harsh. Loudspeakers can also be used as microphones (although the quality will not be as high as that achieved with a properly designed microphone). In this experiment, a loudspeaker will be used as a microphone to inspect the wave-shapes produced by the human voice; wave-shapes that are potentially far more complex than simple sine waves.

25.3 Experimental Example
25.3.1 Equipment
(1) Dual adjustable DC power supply model:_____ S/No.:_____
(1) DMM model:_____ S/No.:_____
(1) Dual channel oscilloscope model:_____ S/No.:_____
(1) Function generator model:_____ S/No.:_____
(2) Small signal NPN transistors (2N3904)
(1) Small signal PNP transistor (2N3906)
(2) Switching diodes (1N914 or 1N4148)
(2) 100 Ω resistor ¼ watt actual: _____
(1) 1 k Ω resistor ¼ watt actual: _____
(1) 6.8 k Ω resistor ¼ watt actual: _____
(1) 1 k Ω potentiometer or decade box
(1) 1 µF capacitor actual: _____
(1) 100 µF capacitor actual: _____
(1) 8 or 16 Ω general purpose loudspeaker

Schematic

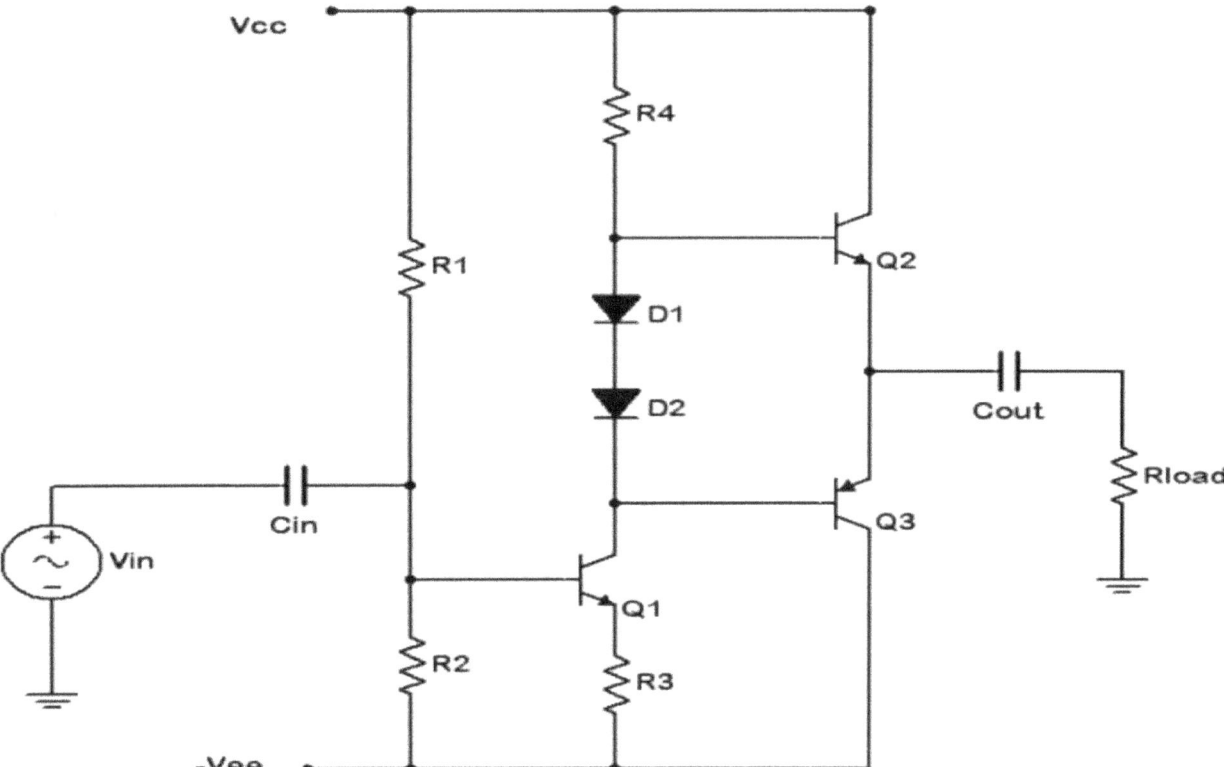

Figure 25.1

25.3.2 Procedure
Bias, Gain and Compliance
1. Consider the circuit of Figure 25.1 using Vcc = 6 volts, Vee = -6 volts, R1 = 6.8 kΩ, R4 = 1 kΩ, R3 = 100 Ω, R_{load} = 100 Ω, Cin = 1 µF and C_{out} = 100 µF. R2 is an adjustable resistance (pot or decade box). For proper bias, the emitters of the output transistors should be at 0 volts DC. For this to be true there must be a voltage of Vcc - Vbe, or approximately 5.3 volts, across R4. Ignoring base currents, this establishes the I_{CQ} of transistor 1 which in turn creates a potential drop across R3. From this the voltage across R2 may be determined. Knowing the value of R1 and the total supply presented, Ohm's law or the voltage divider rule may be used to compute the required setting for R2. Compute the required value for R2 and record it in Table 25.1.
2. Compute the gain of the driver stage. For the load of Q1, the dynamic resistance of the diodes is small enough to ignore. Also, assume the current gain of the output transistors is approximately 100. Remember, only one output transistor is on at any given time. The gain of the class B stage may be assumed to be unity. Record the theoretical circuit gain in Table 25.1.
3. Ideally, the class B stage will produce a compliance of just under 6 volts peak. It may be less than this as the driver stage might clip sooner. Compute the AC load line for the driver stage and determine its compliance. Note that there will be a voltage divider effect between Re and the load of Q1 which will reduce the compliance from that calculated via the load line. Record the theoretical compliance value in Table 25.1. It should be less than that of the output stage and thus represents the compliance of the

entire circuit.

4. Build the circuit of Figure 1 using Vcc = 6 volts, Vee = -6 volts, R1 = 6.8 kΩ, R4 = 1 kΩ, R3 = 100Ω, R_{load} = 100 Ω, Cin = 1 µF and Cout = 100 µF. Set the pot or decade box (R2) to the value calculated in Table 25.1. Disconnect the signal source and inspect the DC voltage at the load. Adjust R2 until this voltage goes to 0 volts. Record the resulting value of R2 in Table 25.1.
5. Connect the signal source and apply a 1 kHz sine at 200 milli-volts peak. Inspect the load and source voltages with the oscilloscope and capture an image of the pair. From these voltages determine the circuit gain and record it in Table 25.1.
6. Increase the signal level until the output begins to clip. Reduce the level until the signal is undistorted and record the resulting load voltage as the experimental compliance in Table 25.1.

Waveforms: Human Perception and Production

7. Turn down the signal source to about 100 mV peak. Insert the loudspeaker in series with the load resistor. Accidentally placing it in parallel will cause excessive current draw and likely destroy the output transistors (after making a particularly loud and irritating squawk). Gradually turn up the signal level while monitoring the load voltage with the oscilloscope. Listen to the sound change as the amplifier begins to clip. Describe this change in Table 25.2. Repeat this with the other frequencies indicated.
8. Remove the loudspeaker and function generator. Reposition the loudspeaker so that it acts as the signal source (i.e., in the original position of the generator). It will now act as a microphone. While examining the load voltage, speak into the loudspeaker and note the typically complex wave shapes. Try holding a few different vowel sounds at different pitches and capture a few of these images. Ordinarily it is difficult for humans to vocalize pure sine waves, however, complex waveforms can be broken down mathematically into a combination of sine waves of differing frequencies, amplitudes and phases. As this is a linear amplifier, superposition holds, and thus if the circuit response to individual sine at differing frequencies can be determined then the response to complex waves such as the human voice and musical instruments can also be determined.

Troubleshooting

9. Remove the loudspeaker and return the generator to the circuit. Consider each of the individual faults listed in Table 3 and estimate the resulting DC and AC load voltages. If the DC voltage moves a great deal off of zero, chances are the AC load voltage will be badly distorted and there is no need to attempt to estimate a precise value. Introduce each of the individual faults in turn and measure and record the load voltages in Table 25.3.

25.3.3 Result Tables
Table 25.1

	Theory	Experimental
R2		
A_v		

Compliance		

Table 25.2

Frequency	Observations
1 kHz	
500 Hz	
200 Hz	

Table 22.5

Issue	$V_{Load\,DC}$	$V_{Load\,AC}$
R_2 Short		
C_{in} Open		
R_1 Open		
R_3 Open		
D_1 Short		
D_2 Open		
C_{out} Open		
V_{CE} Open		

25.3.4 Questions

1. Is the maximum output compliance determined solely by the class B output stage?

2. What kinds of distortion are present in this circuit?

3. Calculate the maximum load power and load current of the amplifier if the loudspeaker had accidentally been placed in parallel with the load resistor rather than in series.

4. How do the values calculated in Question 3 compare to the data sheet maximums for the 2N3904/6?

CHAPTER TWENTY-SIX
JFET Bias

26.1 Objective
The objective of this exercise is to examine three methods to bias JFETs and determine which produce a stable Q point. A method of determining I_{DSS} and $V_{GS(OFF)}$ in the lab is also presented.

26.2 Theory Overview
Unlike bipolar junction transistors, FETs do not have a fixed forward biased junction potential. This makes bias analysis a little trickier. It is often useful to have a couple of device parameters on hand, namely I_{DSS} and $V_{GS(OFF)}$. As is the case with BJTs, finding the main current (I_D) is the key to finding all other circuit currents and voltages. One convenient aspect of JFETs is that the gate current can be ignored for most bias applications. Self Bias may be analyzed through the use of a Self Bias curve or through an iterative process of estimation of V_{GS} leading to drain currents via Ohm's law and the general FET trans-conductance equation. Self Bias tends to have modestly stable Q points. Source Bias is an improvement over Self Bias. It tends to swamp out V_{GS} variation via the addition of a negative source bias voltage. This topology also turns out potentially to have a very stable trans-conductance although it is not examined in this exercise. Finally, Current Source Bias utilizes a BJT to establish a very stable drain current. This turns out this comes at the expense of a stable VGS and trans-conductance (again, not examined here), so this form of bias is not necessarily the best choice for all applications.

26.3 Experimental Example
26.3.1 Equipment
(1) Dual adjustable DC power supply model:_____ S/No.:_____
(1) DMM model:_____ S/No.:_____
(3) Small signal JFETs (MPF102)
(1) Small signal BJT (2N3904)
(1) 2.2 k Ω resistor ¼ watt actual: _____
(2) 4.7 k Ω resistors ¼ watt actual: _____
(1) 330 k Ω resistor ¼ watt actual: _____
MPF102 Datasheet: http://www.onsemi.com/pub_link/Collateral/MPF102-D.PDF

Schematic Diagrams

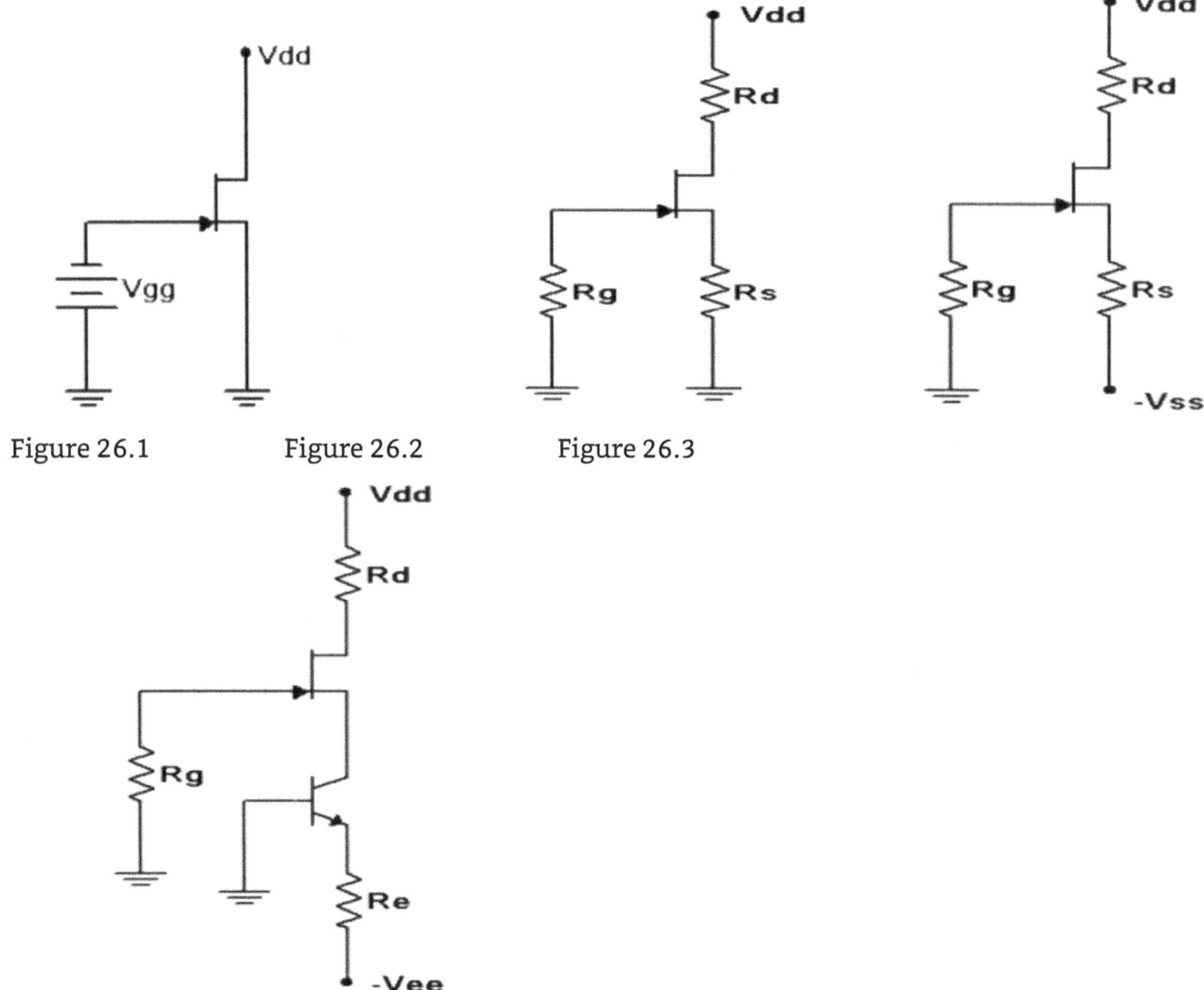

Figure 26.1 Figure 26.2 Figure 26.3

Figure 26.4

26.3.2 Procedure
Determining I_{DSS} and $V_{GS(OFF)}$

1. Consider the circuit of Figure 26.1 using Vdd = 15 volts and Vgg = 0 volts. With nothing else in the circuit, the resulting drain current should equal I_{DSS}. Similarly, if Vgg is gradually changed to a value negative enough to drop the drain current to zero, Vgg must be equal to $V_{GS(OFF)}$.
2. Build the circuit of Figure 26.1 using Vdd = 15 volts and Vgg = 0 volts. Insert an ammeter in the drain and record the resulting current in Table 26.1. Slowly increase the magnitude of Vgg (i.e., make it more negative) until the drain current drops to zero (as a practical point, try to get it under 10 µA, or as low as the ammeter will allow). Record this voltage in Table 1. Repeat this process for the other two transistors. Be sure not to confuse the JFETs. Keep them in order.

Self Bias

3. Consider the circuit of Figure 26.2 using Vdd = 15 volts, Rg = 330 kΩ, Rd = 4.7 kΩ, and

ELECTRONICS LABORATORY MANUAL

Rs =2.2 kΩ. Using the values of Table 26.1, calculate and record the expected voltages for JFET 1 in Table 26.2. Also record the expected drain current in Table 26.3.

4. Build the circuit of Figure 26.2 using Vdd = 15 volts, Rg = 330 kΩ, Rd = 4.7 kΩ, and Rs = 2.2 kΩ. Measure and record the voltages for JFET 1 in Table 26.2. Based on V_D, compute and record the experimental drain current in Table 26.3. Also determine and record the drain current deviation.
5. Repeat steps 2 and 3 for the second and third JFETs.

Source Bias

6. Consider the circuit of Figure 26.3 using Vdd = 15 volts, Vss = -3 volts, Rd = Rs = 4.7 kΩ and Rg = 330 kΩ. A reasonable approximation for V_{GS} in this circuit is -2 volts DC. Based on this, calculate and record the expected voltages for JFET 1 in Table 26.4. Also record the expected drain current in Table 26.5.
7. Build the circuit of Figure 26.3 using Vdd = 15 volts, Vss = -3 volts, Rd = Rs = 4.7 kΩ and Rg = 330 kΩ. Measure and record the voltages for JFET 1 in Table 26.4. Based on V_D, compute and record the experimental drain current in Table 26.4. Also find and record the drain current deviation.
8. Repeat steps 5 and 6 for the second and third JFETs.

Current Source Bias

9. Consider the circuit of Figure 26.4 using Vdd = 15 volts, Vee = -5 volts, Rd = Re = 4.7 kΩ and Rg = 330kΩ. Calculate and record the expected voltages for JFET 1 in Table 26.6. Also record the expected drain current in Table 26.7.
10. Build the circuit of Figure 26.4 using Vdd = 15 volts, Vee = -5 volts, Rd = Re = 4.7 kΩ and Rg = 330 kΩ. Measure and record the voltages for JFET 1 in Table 26.6. Based on VD, compute and record the experimental drain current in Table 26.7. Also find and record the drain current deviation.
11. Repeat steps 8 and 9 for the second and third JFETs.

26.3.3 Result Tables

Table 26.1

JFET	I_{DSS}	$V_{GS(OFF)}$
1		
2		
3		

Table 26.2

JFET	$V_{G\,Theory}$	$V_{S\,Theory}$	$V_{D\,Theory}$	$V_{G\,Exp}$	$V_{S\,Exp}$	$V_{D\,Exp}$
1						
2						
3						

Table 26.3

JFET	$I_{D\,Theory}$	$I_{D\,Experimental}$	% Dev I_D
1			
2			
3			

Table 26.4

JFET	$V_{G\,Theory}$	$V_{S\,Theory}$	$V_{D\,Theory}$	$V_{G\,Exp}$	$V_{S\,Exp}$	$V_{D\,Exp}$
1						
2						
3						

Table 26.5

JFET	$I_{D\,Theory}$	$I_{D\,Experimental}$	% Dev I_D
1			
2			
3			

Table 26.6

JFET	$V_{G\,Theory}$	$V_{S\,Theory}$	$V_{D\,Theory}$	$V_{G\,Exp}$	$V_{S\,Exp}$	$V_{D\,Exp}$
1						
2						
3						

Table 26.7

JFET	$I_{D\,Theory}$	$I_{D\,Experimental}$	% Dev I_D
1			
2			
3			

13.3.4 Questions

1. Of the three biasing forms presented, which produces the most stable and predictable drain current?

   _____
   _____
   _____
   _____

2. Does the precise value of beta for the BJT in the final circuit matter that much? Why/why not?

3. In general, identify two ways of decreasing the drain voltage in the circuit of Figure 26.3.

4. In general, identify two ways of increasing the drain current in the circuit of Figure 26.4.

CHAPTER TWENTY-SEVEN
JFET Amplifiers

27.1 Objective
The objective of this exercise is to examine common source and common drain (voltage follower) JFET amplifiers. Both voltage gain and input impedance will be investigated.

27.2 Theory Overview
In many regards, JFET amplifiers share similar attributes with their bipolar counterparts. Superficially, they look very similar as well. The main functional differences are that JFET based amplifiers tend to have higher input impedances but tend to offer lower voltage gains. Further, without swamping, JFET amplifiers tend to produce lower levels of distortion. As with r'_e impacting bipolar circuit performance, JFET performance is impacted by the trans-conductance, g_m (AKA g_{fs}). Like the bipolar common emitter amplifier, the common source amplifier exhibits a voltage gain greater than one with inversion. The source follower, like the bipolar emitter follower, shows a voltage gain just under one with no inversion.

27.3 Experimental Example
27.3.1 Equipment
(1) Dual adjustable DC power supply model:_____ S/No.:_____
(1) DMM model:_____ S/No.:_____
(1) Dual channel oscilloscope model:_____ S/No.:_____
(1) Function generator model:_____ S/No.:_____
(3) Small signal JFETs (MPF102, substitute J112 if not available)
(2) 4.7 kΩ resistors ¼ watt actual: _____
(1) 22 kΩ resistor ¼ watt actual:_____
(1) 33 kΩ resistor ¼ watt actual:_____
(1) 330 kΩ resistor ¼ watt actual:_____
(2) 10 µF capacitors actual: _____
(1) 470 µF capacitor actual: _____

Schematic diagrams

ELECTRONICS LABORATORY MANUAL

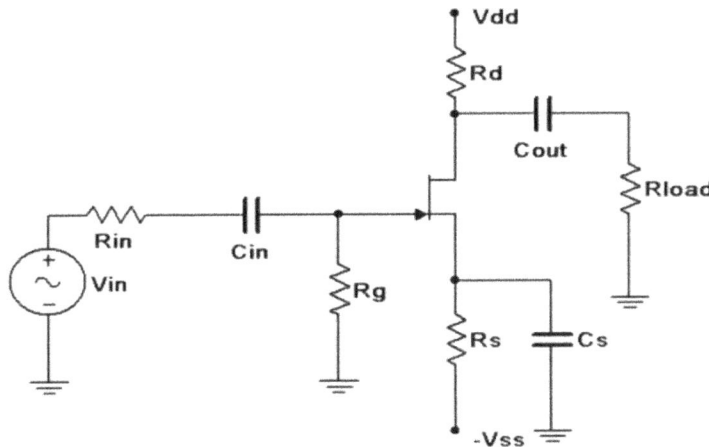

Figure 27.1

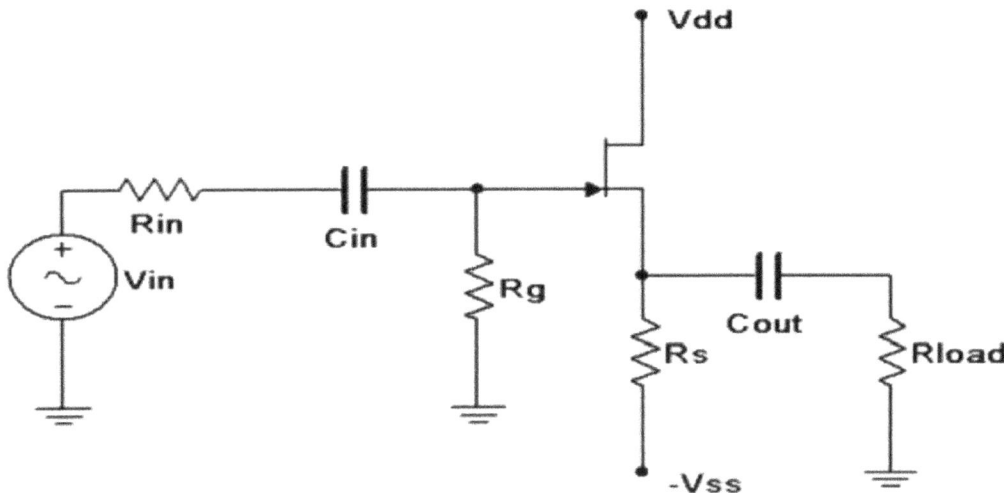

Figure 27.2

27.3.2 Procedure
Common Source Voltage Amplifier

1. Consider the circuit of Figure 27.1 using Vdd = 15 volts, Vss = -3 volts, Rin = 33 kΩ, Rg = 330 kΩ, Rs = 4.7 kΩ, Rd = 4.7 kΩ, R_{load} = 22 kΩ, Cin = Cout = 10 μF and Cs = 470 μF. Assuming V_{GS} = -2 volts and g_m = 2 mS (4 mS if using the J112), determine the theoretical gain and input impedance of the circuit and record these in Table 27.1.
2. Build the circuit of Figure 27.1 using Vdd = 15 volts, Vss = -3 volts, Rin = 33 kΩ, Rg = 330 kΩ,
5. Rs=4.7 kΩ, Rd = 4.7 kΩ, Rload = 22 kΩ, C_{in} = C_{out} = 10 μF and Cs = 470 μF. Set Vin to a 100 mV peak sine at 1 kHz. Measure the voltages at the gate and load, and record these in Table 1. Capture images of the input and gate voltages, and the gate and load voltages. Note whether or not the load is inverted compared to the gate signal.
3. Based on the measured gate and drain voltages, determine the resulting theoretical Av and Z_{in}, and record these in Table 27.1. Note that Z_{in} may be computed using the voltage divider rule or Ohm's law given the gate and input voltages along with the input resistor value. Also determine and record the percent deviations.
4. Repeat steps 1 through 3 for the remaining two JFETs.

Common Drain Voltage Follower

5. Consider the circuit of Figure 27.2 using Vdd = 15 volts, Vss = -3 volts, Rin = 33 kΩ, Rg = 330 kΩ, Rs = 4.7 kΩ, R_{load} = 22 kΩ, Cin = 10 µF and C_{out} = 470 µF. Assuming V_{GS} = -2 volts and g_m = 2 mS (4 mS if using the J112), determine the theoretical gain and input impedance of the circuit and record in Table 27.2.
6. Build the circuit of Figure 27.2 using Vdd = 15 volts, Vss = -3 volts, Rin = 33 kΩ, Rg = 330 kΩ, Rs=4.7 kΩ, R_{load} = 22 kΩ, Cin = 10 µF and C_{out} = 470 µF. Set Vin to a 100 mV peak sine at 1 kHz. Measure the voltages at the gate and load, and record these in Table 2. Capture images of the input and gate voltages, and the gate and load voltages. Note whether or not the load is inverted compared to the gate signal.
7. Based on the measured gate and drain voltages, determine the resulting theoretical Av and Zin, and record these in Table 27.2. Also determine and record the percent deviations.
8. Repeat steps 5 through 7 for the remaining two JFETs.

Troubleshooting

9. Consider each of the individual faults listed in Table 27.3 and estimate the resulting AC load voltage for circuit 1. Introduce each of the individual faults in turn and measure and record the load voltage in Table 27.3.

27.3.3 Result Tables

Table 27.1

JFET	$A_{v\ Theory}$	$Z_{in\ Theory}$	$V_{g\ Exp}$	$V_{d\ Exp}$	$A_{v\ Exp}$	$Z_{in\ Exp}$	% Dev A_v	% Dev Z_{in}
1								
2								
3								

Table 27.2

JFET	$A_{v\ Theory}$	$Z_{in\ Theory}$	$V_{g\ Exp}$	$V_{S\ Exp}$	$A_{v\ Exp}$	$Z_{in\ Exp}$	% Dev A_v	% Dev Z_{in}
1								
2								
3								

Table 27.3

Issue	V_{Load}
R_g Short	
C_{in} Open	
R_d Short	
R_d Open	
R_S Open	
C_{out} Open	

| C_S Open | |
| V_{DS} Open | |

13.3.4 Questions

1. Does the common source amplifier produce a considerable amplification effect and if so, are the results consistent across transistors?

2. Does the common source amplifier produce a phase shift at the load? How does this compare with the common drain follower?

3. How do the voltage gains of these circuits compare to their bipolar versions?

4. How do the input impedances of these circuits compare to their bipolar versions?

CHAPTER TWENTY-EIGHT
JFET Ohmic Region

28.1 Objective
The objective of this exercise is to examine the usage of JFETs in the ohmic region. That is, the device being used as controlled resistance rather than as a current source.

28.2 Theory Overview
For small AC drain-source voltages (<100 mV) the JFET appears as a resistance from drain to source. This resistance is controlled by the DC gate to source voltage. This is referred to as the control voltage or
V_C. The more negative the potential, the larger the resistance. The minimum resistance will be achieved when V_{GS} = 0. If V_{GS} is continuously variable then the JFET behaves as a rheostat, also called a voltage controlled resistor. The addition of a separate resistor in the drain will create a voltage controlled potentiometer. Unlike a true potentiometer, the output signal will not drop to zero due to the minimum on resistance of the JFET ($R_{ds(on)}$). Multiple units can be cascaded for increased attenuation. If the control voltage is set to achieve only the maximum and minimum values, the circuit behaves as a switch that allows or disallows the signal through. This is known as an analog switch.

28.3 Experimental Example
28.3.1 Equipment
 (1) Dual adjustable DC power supply model:_____ S/No.:_____
(1) DMM model:_____ S/No.:_____
(1) Dual channel oscilloscope model:_____ S/No.:_____
(1) Function generator model:_____ S/No.:_____
(2) Small signal JFETs (MPF102, J112)
(2) 4.7 k Ω resistors ¼ watt actual: _____
(1) 10 k Ω resistor ¼ watt actual: _____
J112 Datasheet: https://www.onsemi.com/pub/Collateral/J111-D.PDF
MPF102 Datasheet: http://www.onsemi.com/pub_link/Collateral/MPF102-D.PDF

Schematic Diagrams

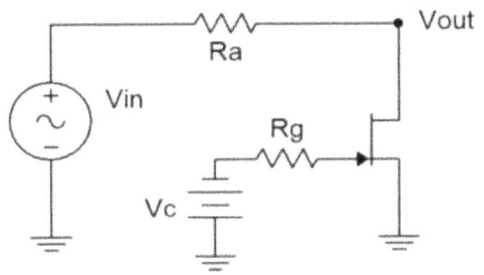
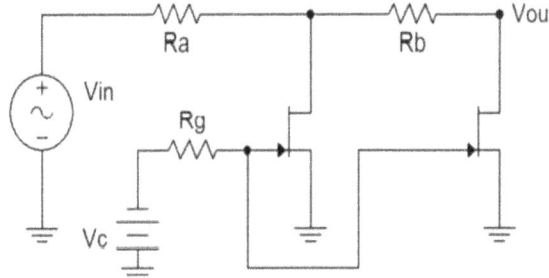

Figure 28.1   Figure 28.2

28.3.2 Procedure
Voltage Controlled Divider
1. Consider the circuit of Figure 28.1 where Vin is a 100 mV peak sine at 1 kHz, Ra = 4.7 kΩ and Rg = 10 kΩ. As Vc varies between 0 and $V_{GS(off)}$, the resistance of the JFET varies between a near open and a minimum value of $R_{ds(on)}$. V_{out} is derived from a voltage divider between Ra and the JFET's resistance.
2. Build the circuit of Figure 1 with Ra = 4.7 kΩ and Rg = 10 kΩ. Set V_{in} to 100 mV peak at 1 kHz. Set the control voltage, V_C, to 0 V_{DC}. Measure the signal at V_{out} using the oscilloscope and record the value in Table 28.1. Also, using the voltage divider rule, determine the effective resistance of the JFET.
3. Repeat step 2 for the remaining control voltages listed in Table 28.1.
4. Using the data from Table 28.1, create a plot of effective resistance versus control voltage.

Analog Switch

5. Build the circuit of Figure 28.2 with Ra = Rb = 4.7 kΩ and Rg = 10 kΩ. Set V_{in} to 100 mV peak at 1 kHz. Set the control voltage, VC, to 0 V_{DC}. Measure the signal at V_{out} using the oscilloscope and record the value in Table 28.2.
6. Set the control voltage to -8 V_{DC}, measure and record the output signal in Table 28.2. Based on these readings, determine the attenuations (V_{out}/V_{in}) and record the results in Table 28.2.

Computer Simulation
7. Repeat steps 2 and 3 using a simulator, recording the results in Table 28.3.

Table 28.2

V_C(DC Volts)	V_{out}	Attenuation
0		
-8		

28.3.3 Result Tables

Table 28.1

V_C(DC Volts)	V_{out}	Effective R_D
0		

-0.5		
-1		
-2		
-3		
-5		

Table 28.3

$V_C(DC\ Volts)$	$V_{out\ Sim}$
0	
0.5	
1	
2	
3	
5	

28.3.4 Questions

1. In Figure 28.1, is the JFET resistance a linear function of control voltage?

2. Detail at least one advantage and one disadvantage of the circuit of Figure 28.1 compared to an ordinary potentiometer.

3. Discuss one advantage of using the circuit of Figure 28.2 instead of a simple mechanical switch.

CHAPTER TWENTY-NINE
Soldering

29.1 Objectives

The objective of this chapter is to examine how to join electronic components to printed circuit board to form a permanent joint, a process called soldering.

29.2 Theory Overview

Soldering is a process in which two or more metal items are joined together by melting and then flowing a filler metal into the joint—the filler metal having a relatively low melting point. It is used to form a permanent connection between electronic components. The metal to be soldered is heated with a soldering iron and then solder is melted into the connection. In this process only the solder melts, not the parts that are being soldered.

Solder is a metallic "glue" that holds the parts together and forms a connection that allows electrical current to flow when it solidifies, see plate 29.1. Solderless breadboard can be used to make test circuits, but to make the circuit durable; soldering the components together becomes necessary. It is different to welding in that the parts being joined are not melted and are usually not the same material as the solder.

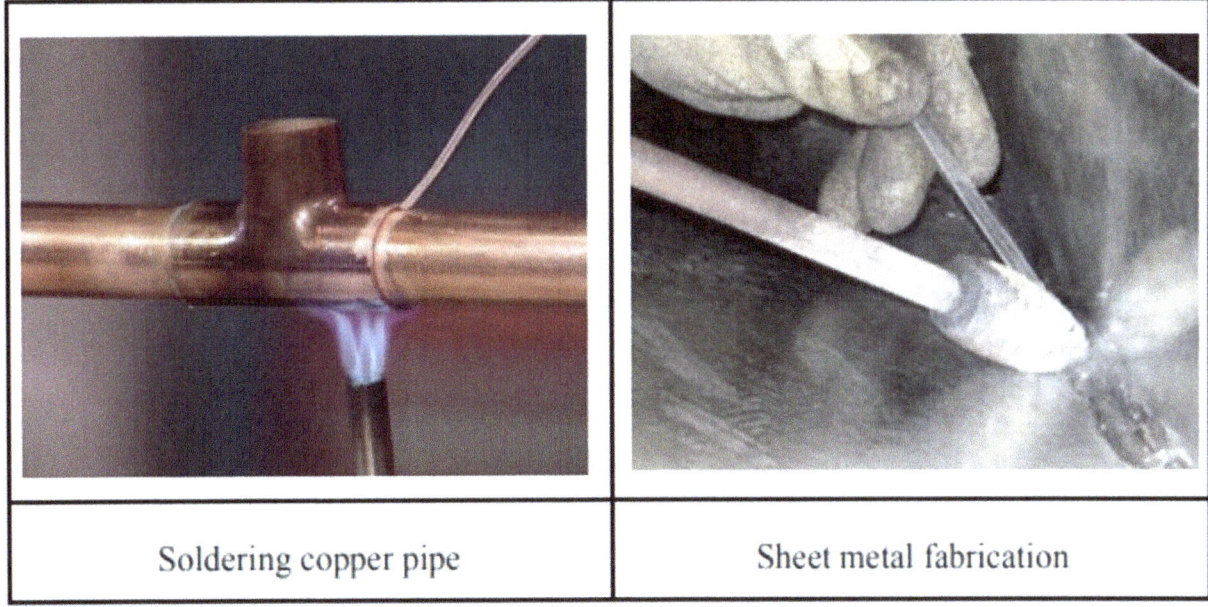

Plate 29.1: Different Types of Soldering

Soldering is a common practice for assembling electrical components and wiring. Although it can be used for plumbing, sheet metal fabrication or automotive radiator repair the techniques and materials used are different to those used for electrical work. This document is intended to provide guidance on the safe working methods and proper tools and techniques for soldering of electrical components.

Soldering Printed Circuit Boards

Soldering may be used to join wires or attached components to a Printed Circuit Board (PCB). Wires, component leads and tracks on circuit boards are mostly made of copper. The copper is usually covered with a thin layer of tin to prevent oxidization and to promote better bonding to other parts with solder. When soldering bare copper wires they are often "tinned" by applying molten solder before making a joint. See plate 29.2.

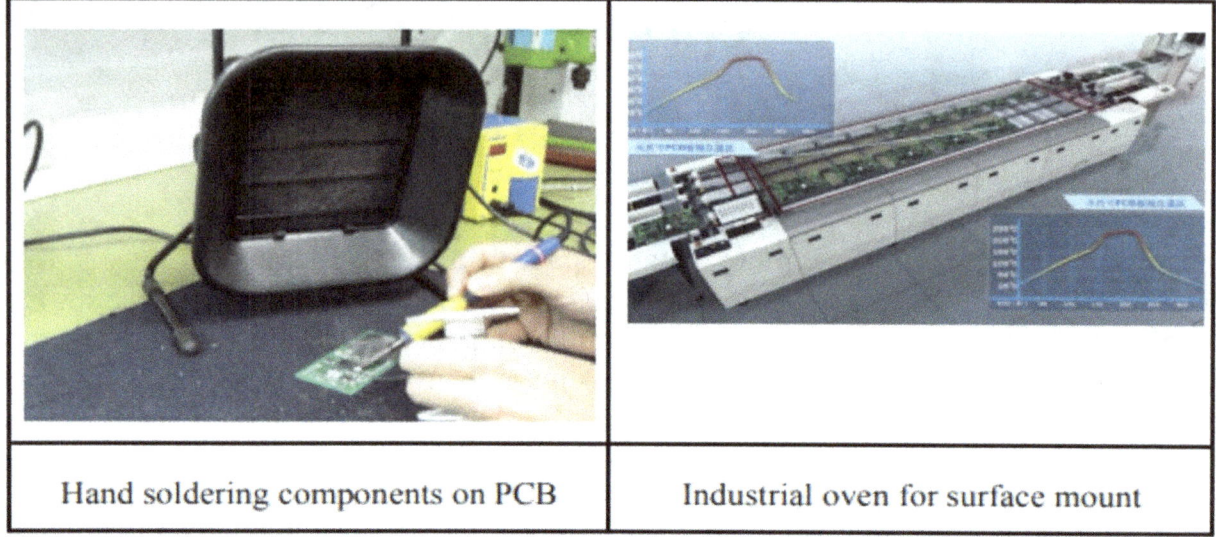

Plate 29.2: – Different Types of PCB Soldering

29.3 Equipments and Materials
29.3.1 Solder

There are different types of solder used for electrical work. They are broadly classified as tin/lead solders or lead free solders. Tin/lead solders have been used for many years because of their ease of use however they have been phased out of commercial use due to the harmful effects on humans and the environment.

Tin/lead solder is still available and is used by "hobbyists" and other noncommercial users as it is still easier to use than lead free types. When using tin/lead (or leaded) solder there are additional safety precautions that must be observed. Plate 29.3 shows different types of solder.

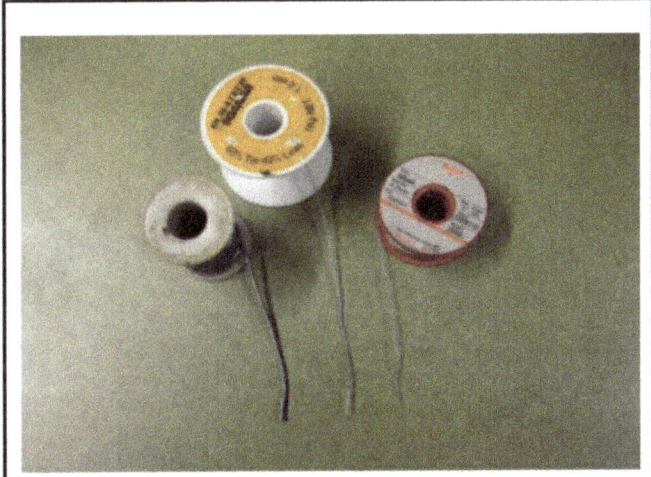	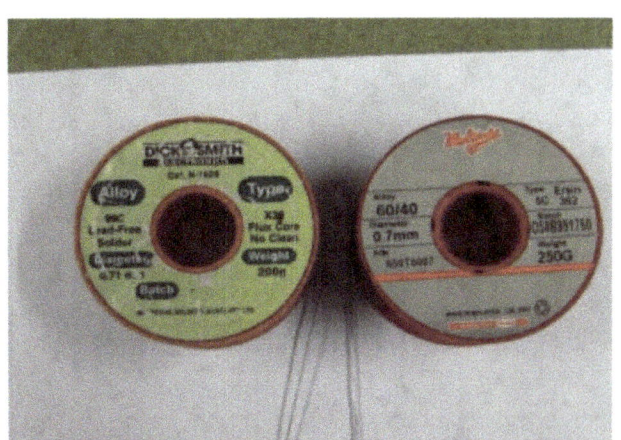
Different gauges of solder wire	Solder composition is labelled (Lead free on left)

Plate 29.3: Different Types of Solder

29.3.2 PCB

Printed circuit boards (PCBs) are populated by electronic components and these may be "surface mount" or "through-hole" types.

Through-Hole Components

As the description "through-hole" suggests, the leads of the component are passed through holes in the PCB and then soldered to a "pad" on the reverse side of the PCB. Soldering is accomplished by heating the component lead and PCB pad with a soldering iron and melting solder wire into the joint. This type of construction was common from the 1960's until early 2000's and is still used by hobbyists and in small scale production where manual assembly is preferred.

Surface Mount Components

Commercial circuits are mostly of the surface mount type as these are cheaper to make, more compact and easier to automate assembly. For surface mount construction the component's pads are on the same side of the PCB as the component and the component connections sit onto these pads. Soldering is accomplished by applying solder paste onto component pads on the PCB, placing the component onto the paste and then heating the entire assembly to melt the solder. Commercial assembly uses ovens to heat the boards. Hobbyists can also use surface mount components and soldering can be accomplished by applying solder paste and melting with a hot plate, small oven or soldering iron. Some surface mount joints can be soldered using a soldering iron and solder wire. See plate 29.4.

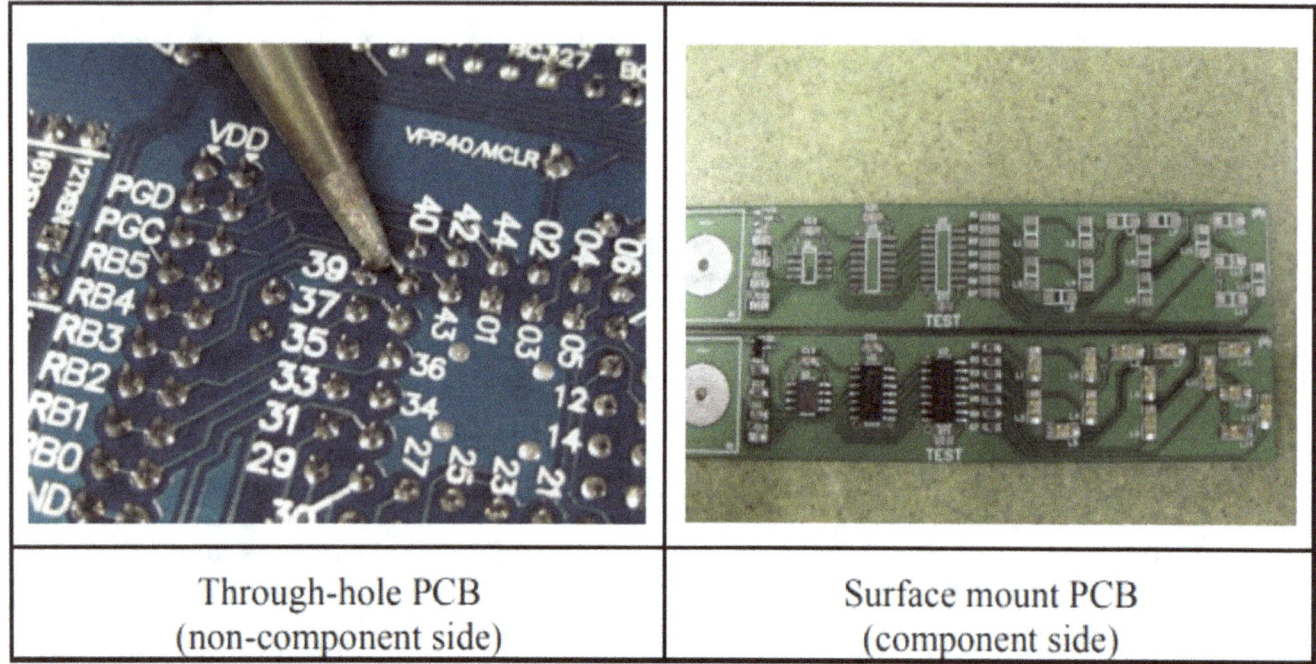

Plate 29.4: Different Types of PCB

29.3.3 Flux

For electrical soldering both solder wire and solder paste contain flux. This helps to clean the surfaces being soldered and prevent oxidization of the hot solder. The composition of the flux will vary depending on whether it is in a paste or wire, leaded or unleaded solder. Solder wire usually contains a flux called "rosin". Most fluxes will produce fumes when the solder is heated and these fumes are likely harmful to your health. For occasional soldering it may be sufficient to have a well-ventilated workspace but for longer or repeated exposure a fume extractor should be used. Solder flux can also cause solder to spatter and eye protection should be worn when soldering. Plate 20.5 shows fumes and fume extraction system.

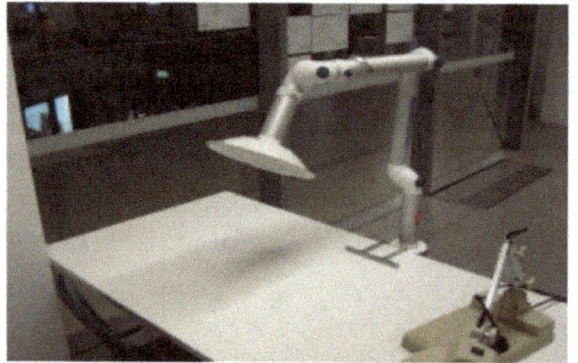

Plate 29.5: Fumes and Fume Extraction Systems

29.3.4 Soldering Irons

Soldering irons come in many varieties and sizes. Soldering irons may be electric, gas powered or externally heated. Most common types are electric. Simple electric soldering irons have no controls and you simply plug them in and wait for them to heat up. Their temperature is regulated by the power of the heating element and heat loss to the environment. Some soldering irons have temperature controls which allow the user to set a desired operating temperature for the soldering iron. This is useful if the soldering iron is being used for different types of solders which have different melting points or if the soldering iron is being used for other purposes such as heating. It also introduces a problem if the user does not set an appropriate temperature for the work; solder can be overheated and decompose. Hotter is not better! A temperature of around 320 °C works well for 60/40 leaded solder. Some temperature controlled soldering irons use interchangeable tips to change the temperature at which they operate. See plate 29.6 for different types of soldering irons, and plate 29.7 for different types of workstation.

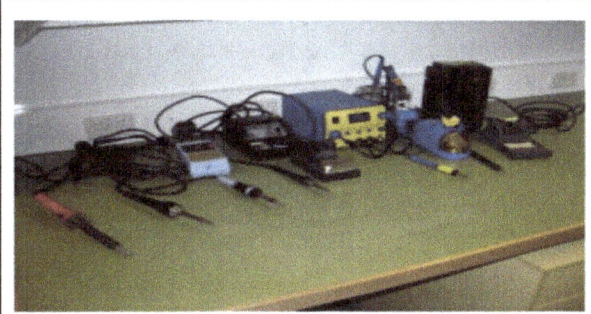 

Plate 29.6: Types of soldering irons

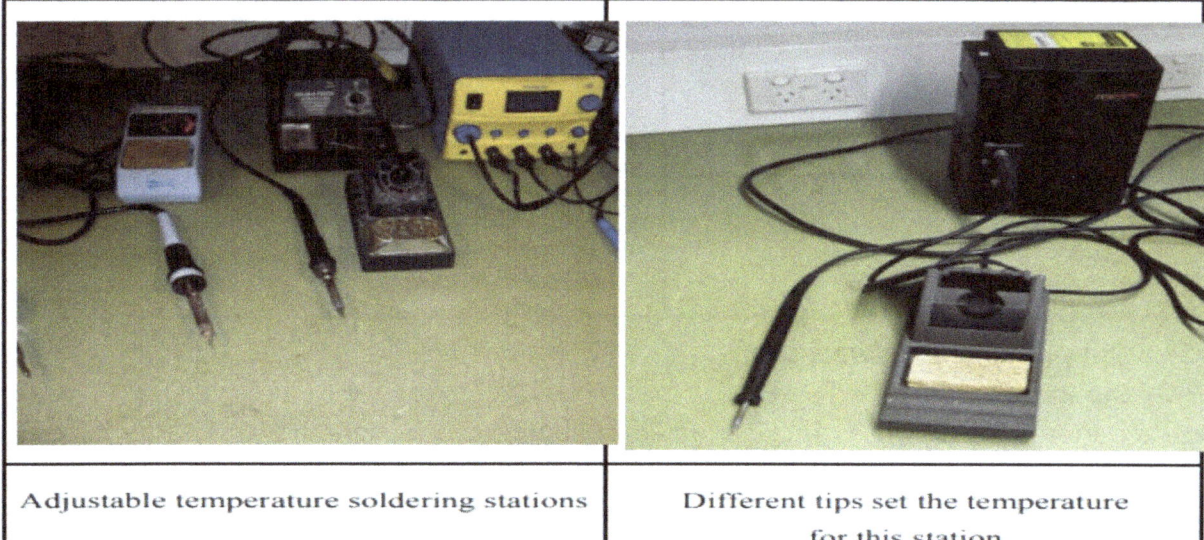

Plate 29.7: Types of soldering stations

29.3.5 Tips

Heat is transferred from the tip of the soldering iron to the joint by thermal conduction enabled by metal to metal contact between the tip and joint. The tips of soldering irons come in various shapes and sizes to enable the best contact to be made. Most tips are either conical or chisel shapes. The shape is largely a personal preference and you can use whichever type works best for you. The size of the tip should be selected to allow the tip to be placed against the joint being soldered without interfering with adjacent parts. The tip should be large enough to conduct sufficient heat into the joint to allow the solder to melt and flow properly. The choice of tip size is not a precise calculation and a "normal" size tip will work for most joints on a PCB.

29.3.6 Tip Contamination and Cleaning

The thermal conduction from the tip to the joint may be inhibited by contamination on the tip. This contamination can be formed by burnt solder flux or oxidized solder. To make best thermal contact the tip should be cleaned using a tip cleaner(!). Two types are a wet sponge or a brass wire wool. The wet sponge removes the contaminated material when the tip is wiped across it, the water in the sponge cools the solder and the mechanical abrasion removes the contamination leaving a thin coating of clean solder on the tip. This method can cause the tip temperature to dip momentarily. The brass wire wool type removes the contamination by mechanical abrasion and bonding contaminated solder to the brass. The tip is pushed into the brass wool and when it is withdrawn the tip is clean with a thin coating of solder. You must not "wipe" the tip on the brass wool type because the springiness of the brass wool may flick molten solder which may cause burns to people or objects. You should never "flick" excess solder from the soldering iron as this may also cause burns or damage. Plate 29.8 shows dirty tip and brass wool cleaner.

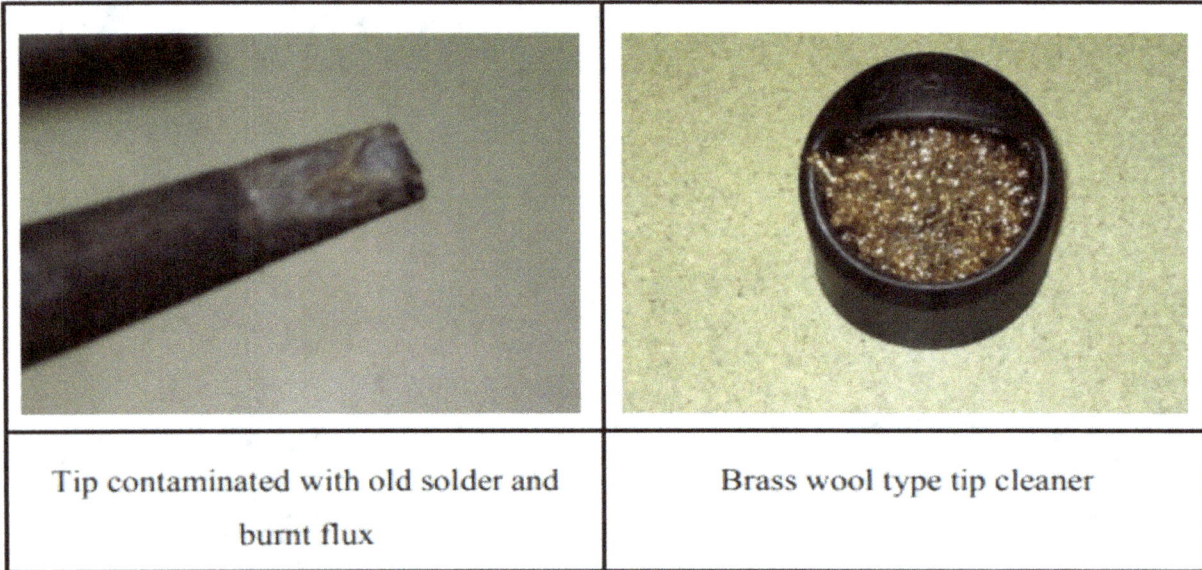

Plate 29.8 Dirty tip and brass wool cleaner.

29.3.7 Ovens and Hot Plates

For surface mount soldering the heat is usually applied to the whole PCB and all components soldered at the same time. For commercial work this is done in large ovens, often with conveyors to move the boards through the oven. For small scale work simple infrared ovens are available. Another technique uses a hotplate. The same considerations for temperature apply, although because the heat source does not come into contact with the solder or flux, contamination is less likely. Time and temperature are considerations with these methods as the components are exposed to the high temperature for the period required for the solder paste to melt and flow. Plate 29.9 shows oven and hot plate used for surface mount soldering

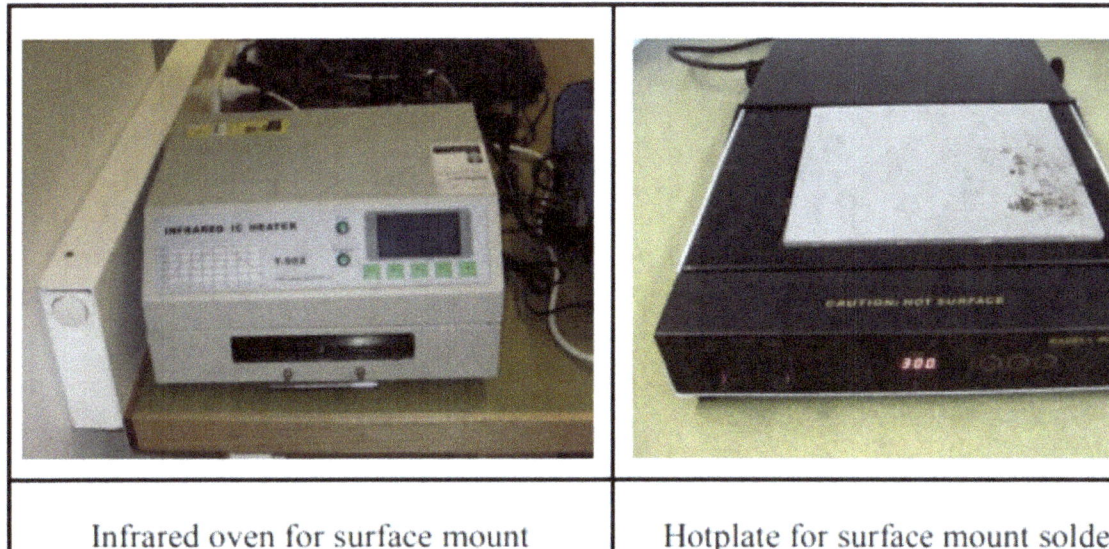

Plate 29.9: Oven and hot plate used for surface mount soldering

29.3.8 Desoldering

If a part that has been soldered needs to be replaced it needs to be "de-soldered". Depending on the part and type of joint it may be possible to simply re-melt the solder and remove the part, or it may be necessary to remove the solder from the joint so the part can be freed. Some methods for removing solder are solder wick, solder sucker or de-soldering tool. Solder wick is a copper braid which is applied to the joint and heated with a soldering iron. As the solder in the joint is melted it is drawn into the solder wick like a sponge and is removed from the joint. A solder sucker is a spring loaded syringe or rubber bulb. The tip of the solder sucker is placed near the joint as the joint is melted by a soldering iron. When the sucker is operated a vacuum is created which draws the molten solder from the joint into the body of the sucker. A de-soldering tool is a type of soldering iron with a hollow tip and is connected to a pump or vacuum source. The tip of the de-soldering tool is placed onto the joint, typically over a component lead, and once the solder has melted the pump is operated to draw the molten solder away. Plate 29.10 shows desoldering tools and tweezers for surface mount devices.

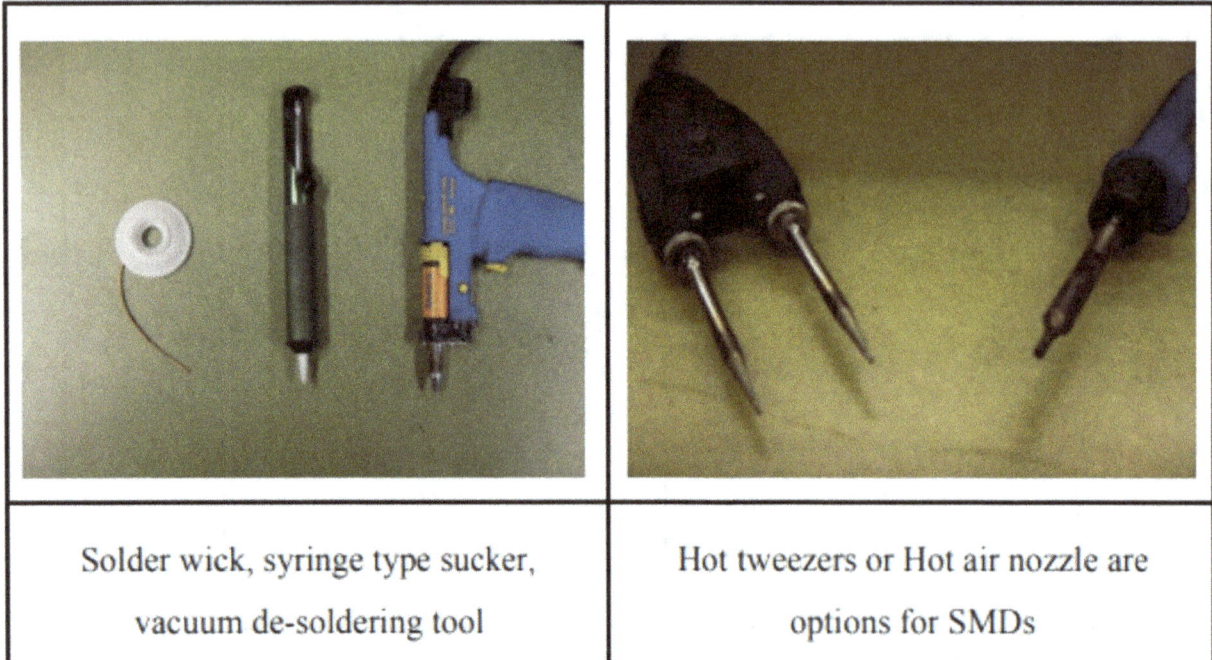

Plate 29.10 Desoldering tools and tweezers for surface mount devices.

29.4 Hazards involved in soldering
29.4.1 Heat
Although solder has a relatively low melting point this temperature is more than high enough to cause serious burns to people or objects. It is important to hold the soldering iron only by the insulated handle, never touch the heating element or tip when the soldering iron is on. The soldering iron will remain hot for some time after it is turned off so always check that it has cooled down before touching it, e.g. if changing the tip. When you are not soldering always keep the soldering iron in a proper holder so that you don't touch it accidently and it doesn't heat or burn other objects such as the bench-top. Don't hold parts being soldered with your hands as these will also be heated when being soldered. Don't flick molten solder from the soldering iron or wipe the tip on brass wool type tip cleaners.

If using a hot plate for surface mount soldering do not touch the hot plate. Use utensils such as pliers to place and remove PCB's from the hotplate. If using hot air tools for soldering, de-soldering or rework, do not direct the hot air stream onto yourself or other people. If using an oven allow the PCB to cool before handling or use utensils. Don't place hot PCB's on temperature sensitive surfaces. If burns occur they should be treated by holding under cold running water for several minutes and assistance sought if burns are severe. Incidents should be reported.

29.4.2 Toxic materials
Leaded solder contains lead which is a harmful material. Use of this type of solder will probably involve handling it and your skin may become contaminated by it. Although it is unlikely that the lead can be absorbed directly through your skin it may be ingested indirectly if it is transferred by handling food whilst your skin is contaminated. Always wash your hands thoroughly before eating or handling food.

Solder flux creates fumes when heated during soldering which may be harmful if inhaled. Use a fume extractor to avoid inhaling fumes.

29.4.3 Spattering

Solder and flux can spit or spatter when heated. Always wear eye protection (safety glasses) when soldering.

29.4.3 Electrical Safety

Electric soldering irons are plugin appliances and must have a current safety test tag. The test will confirm that the soldering iron conforms to electrical safety standards and has not been damaged at the time of the test. Before use you should visually check that the soldering iron does not have damage such as melted insulation on the lead, broken or cracked handle or exposed conductors. Don't use damaged equipment and report the damage. For electrical safety the exposed metal parts such as the tip and heating element are earthed. Don't solder on any live equipment as contact with the earthed tip may cause damage to the equipment or soldering iron.

CHAPTER THIRTY
Soldering Practice

30.1 Objective
To learn how to solder circuit components on Printed Circuit Board (PCB)

WARNINGS!!!

This lab involves hot soldering irons and components. **USE EXTREME CAUTION!!!** Any reckless actions will NOT be tolerated.

a) This is a general procedure for performance typical soldering or de-soldering operations. It does not apply to specialized soldering processes or equipment.
b) Make sure you fully understand how to use the soldering iron and examine the tools before use. Ask for help if you have any concerns or questions related to the tools.
c) Only set the irons at the proper temperature range when in use. Switch off and/or unplug soldering tools when not in use. Allow soldering tool to cool before storing.
d) Beware of hot soldering irons and components. Make sure others are aware of hot tools as well. Make sure you place the hot iron in an appropriate holder to prevent heat or fire damage, even if they are cooling.
e) Some solders used in laboratories may contain lead. DO NOT breathe fumes generated while soldering.
f) Wash your hands after using solder and soldering tools.

30.2 Theory Overview
30.2.1 Properties of Solder
The solder used for electronics is a metal alloy, made by combining tin and lead in different proportions. You can usually find these proportions marked on the various types of solders.
With most tin/lead solder combinations, melting does not take place all at once. Fifty-fifty solder begins to melt at 183 C (361 F), but it is not fully melted until the temperature reaches 216 C (420 F). Between these two temperatures, the solder exists in a plastic or semi-liquid state. The plastic range of a solder varies, depending upon the ratio of tin to lead. With 60/40 solder, the range is much smaller than it is for 50/50 solder. The 63/37 ratio, known as eutectic solder has practically no plastic range, and melts almost instantly at 183 C (361 F).

30.2.2 Flux
Reliable solder connections can only be accomplished with truly cleaned surfaces. Solvents can be used to clean the surfaces prior to soldering but are insufficient due to the extremely rapid rate at which oxides form on the surface of heated metals. To overcome this oxide film, it becomes necessary in electronic soldering to use materials called fluxes. Fluxes consist of natural or synthetic rosins and sometimes-chemical additives called activators to remove oxides and keep them removed during the soldering operation. The flux action is very corrosive at solder melt

temperatures and accounts for removing metal oxides. In its unheated state, however, rosin flux is non-corrosive and non-conductive and thus will not affect the circuitry.

30.2.3 Controlling Heat
Although controlling soldering iron tip temperature is not the key element in soldering, the advised soldering iron tip temperature is in range of 650 ~ 750 F. More importantly, the key element is controlling the heat cycle of the work, i.e. how fast the work gets hot, how hot it gets, and how long it stays hot is the element to control for reliable solder connections.

30.2.4 Thermal Mass
Each joint, has its own particular thermal mass, and how this combined mass compares with the mass of the iron tip determines the time and temperature rise of the work.

30.2.5 Surface Condition
A second factor of importance when soldering is the surface condition. If there are any oxides or other contaminants covering the pads or leads, there will be a barrier to the flow of heat. Even though the iron tip is the right size and temperature, it may not be able to supply enough heat to the joint to melt the solder.

30.2.6 Thermal Linkage
This is the area of contact between the iron tip and the work. Figure 30.1 shows a view of a soldering iron tip soldering a component lead. Heat is transferred through the small contact area between the soldering iron tip and pad. The thermal linkage area is small.

Figure 30.2 also shows a view of a soldering iron tip soldering a component lead. In this case, the contact area is greatly increased by having a small amount of solder at the point of contact. The tip is also in contact with both the pad and component further improving the thermal linkage. This solder bridge provides thermal linkage and assures the rapid transfer of heat into the work.

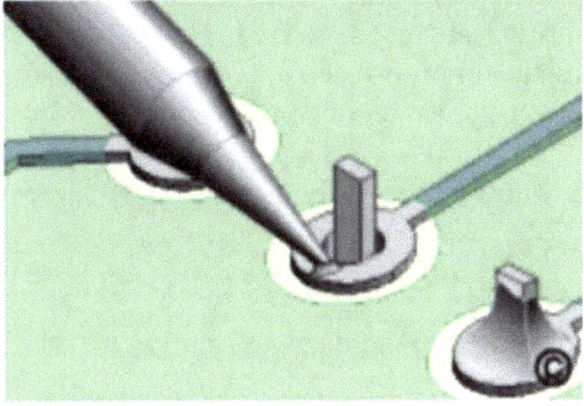 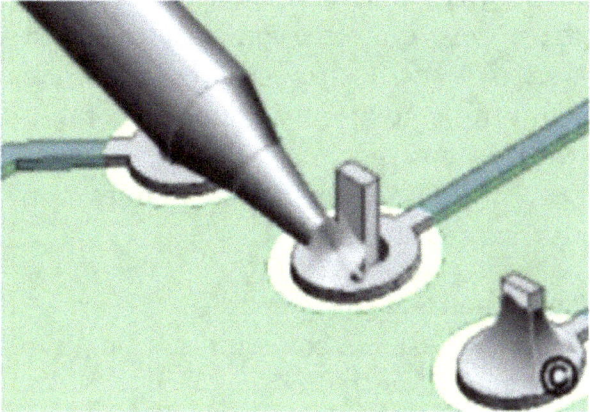

Figure 30.1: Minimal thermal linkage due to insufficient solder between the pad and soldering iron tip.

Figure 30.2: A solder bridge provides thermal linkage to transfer heat into the pad and component lead.

30.2.7 Re-soldering
Care should be taken to avoid the need for re-soldering. When re-soldering is required, quality standards for the re-soldered connection should be the same as for the original connection.

30.2.8 Workmanship

Solder joints should have a smooth appearance. A satin luster is permissible. The joints should be free from scratches, sharp edges, grittiness, looseness, blistering, or other evidence of poor workmanship. Probe marks from test pins are acceptable providing that they do not affect the integrity of the solder joint. An acceptable solder connection should indicate evidence of wetting and adherence when the solder blends to the soldered surface. The solder should form a small contact angle; this indicates the presence of a metallurgical bond and metallic continuity from solder to surface. See Figure 30.3.

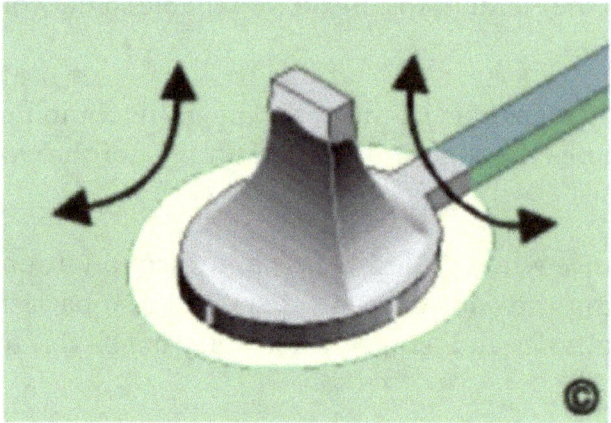

Figure 30.3: Solder blends to the soldered surface, forming a small contact angle

30.2.9 Lab Deliveries
PRELAB:
Please watch the following videos read up the blog.
a) https://www.youtube.com/watch?v=Qps9woUGkvI
b) https://www.youtube.com/watch?v=fYz5nIHH0iY
c) EEVblog #186 - Soldering Tutorial Part 3 - Surface Mount - YouTube
d) https://www.youtube.com/results?search_query=universal+pcb+soldering
e) https://www.electronicsandyou.com/blog/smd-surface-mount-electronic-components-for-smt.html#:~:text=Availability%20of%20Different%20Types%20of%20SMD%20Components
f) The PCB Soldering Techniques You Should Know - Free Online PCB CAD Library (ultralibrarian.com)

30.3 Components & Equipment
- HAKKO FX-888D soldering station (i.e. power, soldering iron, holder, wet sponge)
- Solder wire with flux
- Wire cutter, PCB and circuit components/parts (e.g. resistors, capacitors, LEDs, etc.)
- Other accessories (if possible): helping hands flux cleaner, safety goggles, desk fan, etc.

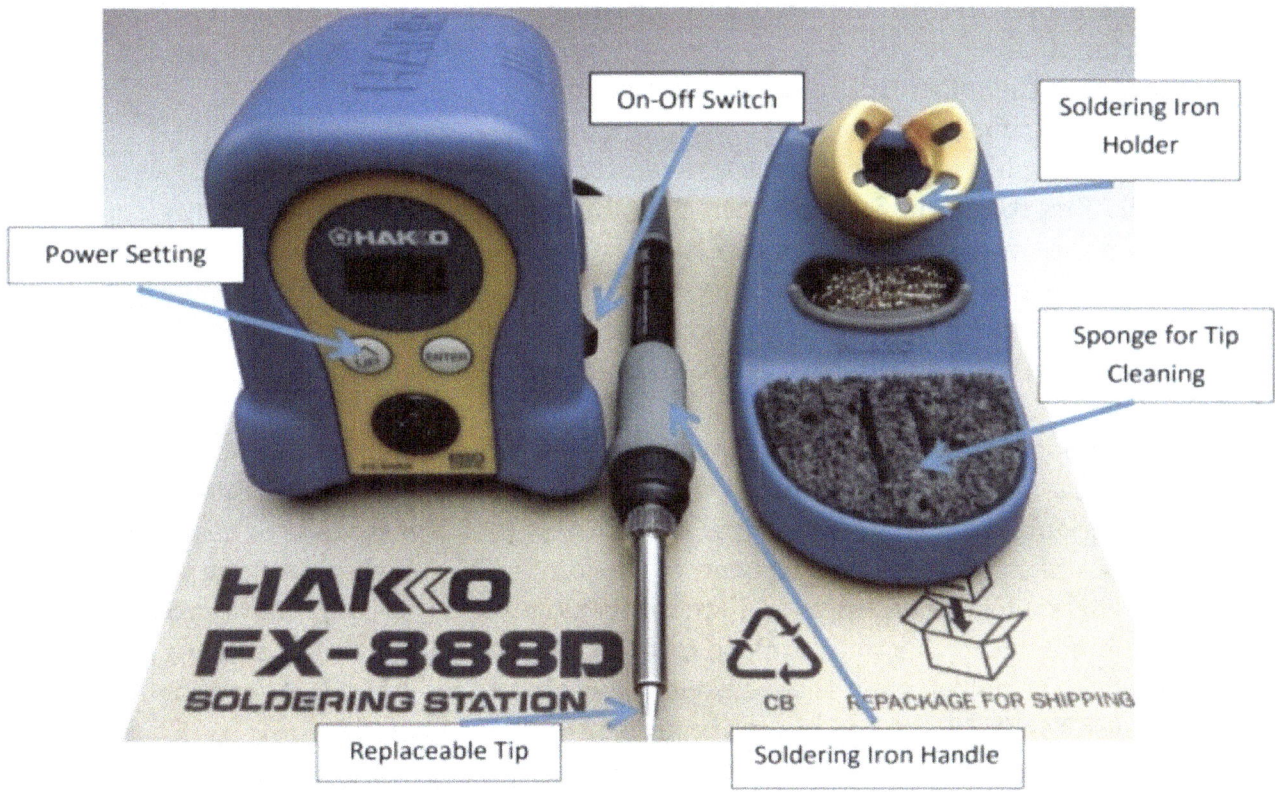

Plate 30.1: Soldering Station

30.4 Lab Experiments:
30.4.1. Setting Up Work Area.
  a. Ensure adequate ventilation. If multiple people are soldering in a concentrated area, set up a fan to gently blow fumes and vapors away from you and your co-workers.
  b. Keep area clean around workplace at all times.
  c. When working with statically sensitive components (most active devices such as ICs, FETs, transistors, etc.), be sure to use an anti-static mat to work on and wear an anti-static wrist strap to minimize risk of electrostatic discharge (ESD) damage.

30.4.2. Preparing Workpiece.
  a. Place the PCB steady on the desk, or clamp work securely while performing soldering or desoldering. When necessary, use a vise, a helping hand or other approved clamping systems to keep your hands free to work.
  b. Use heat sinks to protect thermally-sensitive circuit components.
  c. When soldering wire connections, make sure the wires are tightly connected. Use appropriate covering like heat shrink tubing or twist-on connectors to protect the splice. Do not use wires with melted insulation or exposed conductors.

30.4.3. Setting Up the Soldering Station.
  a. Select proper solder and flux. Most solder nowadays are integrated with flux.
  b. Select the proper sized solder tip for your work. As trace and pad size decrease, soldering tip size must also decrease.
  c. Turn the soldering station ON and set the temperature 650~750 F. Note: higher

temperatures lead to more rapid formulation of oxidation on soldering tip and will shorten tip life.

d. Make sure the solder tip is cleaned and tinned.

30.4.4. Tinning Soldering Tip
a. Allow the soldering iron to reach temperature.
b. Apply flux to the tip first and then liberally apply solder to tip. Note: flux-core solder may not require application of flux.
c. Wipe off excess solder on sponge.
d. A properly tinned tip will be shiny and free of oxidation.
e. It is normal to have to re-tin a tip from time to time as oxidation builds up on the tip.

30.4.5. Soldering
a. Hold the soldering iron like a pen, near the base of the handle.
b. Touch the soldering iron onto the joint to be made. b. Make sure it touches both the component lead and the track. Hold the tip there for a few seconds to heat the joint.
c. Unroll the solder and bring the end to the joint to be soldered near the soldering iron tip. The solder should melt and smoothly flow onto the components to be soldered (component lead, pcb trace, etc). Be sure to apply the solder to the joint, not the iron.
d. Remove the solder, then the iron, while keeping the joint still. Allow the joint a few seconds to cool before you move the circuit board.
e. Inspect the joint closely. Be sure that the solder joint is good, as described below.

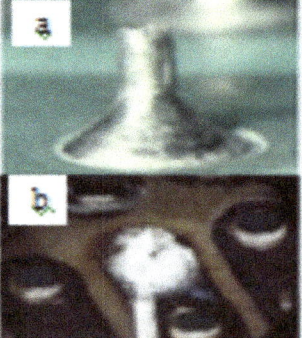

f. Solder joints may be cleaned after they have cooled using isopropyl alcohol and a Q-tip or similar cotton swab.

30.4.6. Solder Joint Evaluation
a. **Good Solder Joints**
- will be shiny and concave in nature.
- will be clean and free of dirt and voids.
- will fully cover the pad the component lead lays on or passing through.

b. **Cold Solder Joints**
- do not make a good electrical or mechanical connection
- can be remedied by removing existing solder (de-soldering and clean-up), and then re-soldered.

30.4.7. De-soldering
De-soldering is done to remove components that have been soldered together. Two common forms of de-soldering are done through the use of a vacuum plunger device (solder sucker) or by applying a braided wick to the joint that pulls solder away through capillary action.

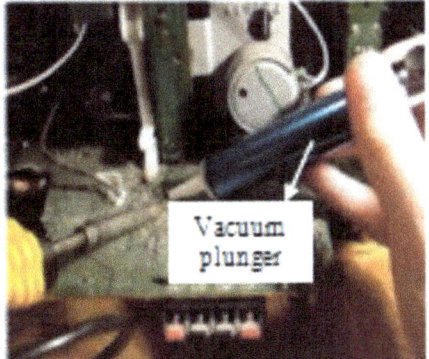

a. Use of Vacuum Plunger

- To use a vacuum plunger, cock the plunger by depressing it fully until it latches.
- Bring the vacuum plunger to the location of the joint to be de-soldered.
- Heat the joint with the tip of the soldering iron.
- When the solder has melted, press the trigger of the device, which will pull the solder out of the joint.
- Two or more tries may be needed to fully remove the solder from the joint.
- On occasion, it is actually helpful to add solder to the joint to provide addition thermal mass and uniform heating to the joint. This can assist in complete melting of the solder in the joint.

b. Use of Solder Wick

- Solder wick is typically a ribbon of braided fine copper wire with rosin core flux impregnated.
- To use solder wick, lay the wick over the joint to be de-soldered.
- Apply the heated tip of the soldering iron to allow the wick to be heated and melt the solder in the joint.
- The solder will flow out of the joint and into the wick through capillary action.

30.5 Postlab Report:

Include the following elements in the report document:

1. Theory of operation Include a brief description of every element and phenomenon that appears during the experiments.
2. Prelab report: None. Read through Background and watch the linked videos in the Prelab.
3. Results of the experiments Number of Experiments: 1 Experiment Results: Photos of soldered parts on PCB
4. Answer the questions
   1) What tools do you need to solder electrical/electronic parts onto a PCB?
   2) What is the temperature range for a soldering iron to work properly?
   3) What precautions need to be taken during the soldering process?
   4) What is the most important thing during soldering?

5. Conclusions Write down your conclusions, things learned, problems encountered

during the lab and how they were solved, etc.
6. Images Paste images (e.g. scratches, drafts, screenshots, photos, etc.) in Postlab report document (only .docx, .doc or .pdf format is accepted). If the sizes of images are too large, convert them to jpg/jpeg format first, and then paste them in the document.

Attachments (If needed) Zip your projects. Send as attachments to the tutor's email address, or provide link to the zip file on Google Drive / Dropbox, etc.

APPENDIX A: CREATING GRAPHS USING A SPREADSHEET

While nothing beats good data plotting and analysis software (check out SciDAVis for an excellent free program), you can also create a variety of graphs using spreadsheets such as the one in Open Office or
Excel (Microsoft Office). What follows works for Excel 2007 and Open Office 4. Other versions may have different menus and options. Here's how to take your tabular data from lab and create a graph. These instructions assume you will set the independent axis on the horizontal and the dependent axis as the vertical. This is the typical case but there are exceptions (see note at end). Remember, the independent axis presents the input parameter you set (e.g., a power supply voltage or a mass) and the dependent axis presents the output parameter (i.e., the item you are interested in and have measured as an outcome such as a resulting current or change in position).

1. Open a new worksheet. In the first column (column A), enter the text for the legend. This is particularly important if you're plotting multiple datasets on a single graph. Starting in the second column (column B), enter values for the horizontal (independent) axis on the first row of the worksheet. In like fashion, enter values for the vertical (dependent) axis on the second row. For multiple trials, enter the values on subsequent rows. For example, if you are setting a series of voltages in a circuit and then measuring the resulting currents, the voltages would be in row one and the currents in row two. If you changed the circuit components, reset the voltages, remeasured the currents and wish to compare the two trials, then the new set of currents would be in row three and so on. Each of these rows would have their identifying legend in column A with the numeric data starting in column B. Specifically, the legend text for the first data set would be in cell $A$2 and the numeric values would be in cells $B$2 through $X$2 (where X is the final data column), for the second set the legend text would be in cell $A$3 and the numeric values would be in cells $B$3 through $X$3, etc.
2. Select/highlight all of the data (click the first cell, in the upper left corner, and drag the mouse over all of the cells used).
3. Select the Insert menu and choose Chart. Ordinarily you will use an XY Scatter chart. There are other options but this is the one you'll need in most cases. A simple Line chart is not appropriate in most cases. You might get a graph that "sort of" looks correct but the horizontal axis will simply represent the measurement sequence (first, second, third) rather than the value you set.
4. You can customize the appearance of the chart. In general, you can edit items by simply double-clicking on the item or by using a right-mouse click to bring up a property menu. This will allow you to add or alter grid lines, axes, etc. You can also stipulate variations such as using data smoothing, adding a trend line, etc. It is possible to

change the axes to logarithmic or alter their range; and fonts, colors and a variety of secondary characteristics may be altered.

5. Once your chart is completed, you may wish to save the worksheet for future reference. To insert the chart into a lab report, select the chart by clicking on it, copy it to the clipboard (Ctrl+C), select the insertion point in the lab report, and paste (Ctrl+V).

6. In those odd instances where you need to reverse the dependent and independent axes such as a VI plot of a diode where currents are set and resulting voltages are measured, but you want the voltage on the horizontal; some spreadsheets have an axis swap function. If not, you'll need to swap the data ranges for the chart axes. For example, following the instructions above, your independent/horizontal axis is row one. The data are in cells $B$1 through $X$1. The dependent data are in cells $B$2 through $X$2. These ranges can be seen in the chart's Data Series or Data Range menu or dialog box. It will say something like: "X Values: =Sheet1!$B$1:$F$1" and "Y Values: =Sheet1!$B$2:$F$2". Simply swap the row numbers so that it says "X Values: =Sheet1!$B$2:$F$2" and "Y Values: =Sheet1!$B$1:$F$1".

7. Data smoothing can be useful to remove the "jaggyness" of some plots. For simple curves, a second degree B-Spline is suggested if you're using Open Office. For data that are expected to be linear, a trend line can be useful to better see the approximation.

Here is an example worksheet showing a plot of two resistors. The first plot is basic, the second uses smoothed data with a linear trend line:

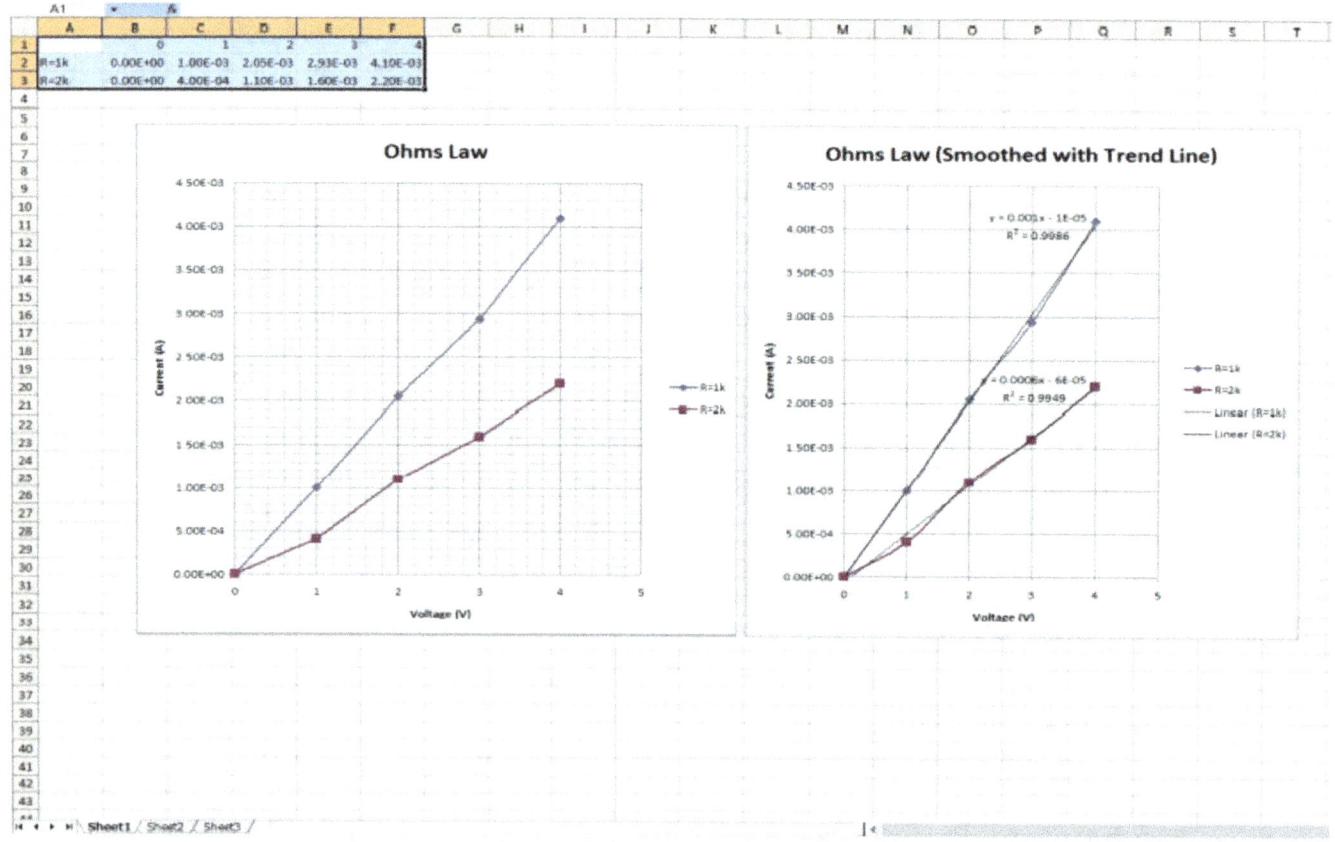

APPENDIX B: MANUFACTURER'S DATASHEET LINKS

Diodes
1N4002 Datasheet: https://www.onsemi.com/pub/Collateral/1N4001-D.PDF
1N4148 Datasheet: https://www.onsemi.com/pub/Collateral/1N914A-D.pdf
1N751 Datasheet: http://www.digitroncorp.com/Documents/Datasheets/1N746-1N759A,-1N4370-1N4372A.aspx?ext=.pdf
1N914 Datasheet: https://www.onsemi.com/pub/Collateral/1N914A-D.pdf
NZX5V1B Datasheet: https://assets.nexperia.com/documents/data-sheet/NZX_SER.pdf

Standard Red LED Datasheet:
https://www.sparkfun.com/datasheets/Components/LED/COM-09590-YSL-R531R3D-D2.pdf

High Brightness White LED Datasheet:
http://cdn.sparkfun.com/datasheets/Components/General/YSL-R1042WC-D15.pdf
IR Detector Datasheet: http://optoelectronics.liteon.com/upload/download/DS-50-93-0013/LTR-301.pdf
IR Emitter Datasheet: http://optoelectronics.liteon.com/upload/download/DS-50-92-0009/E302.pdf

Transistors
2N3904 Datasheet: https://www.onsemi.com/pub/Collateral/2N3903-D.PDF
2N3906 Datasheet: https://www.onsemi.com/pub/Collateral/2N3906-D.PDF
J112 Datasheet: https://www.onsemi.com/pub/Collateral/J111-D.PDF
MPF102 Datasheet: http://www.onsemi.com/pub_link/Collateral/MPF102-D.PDF

Miscellaneous
GL5528 CdS Cell Datasheet: http://cdn.sparkfun.com/datasheets/Sensors/LightImaging/SEN-09088.pdf
Vishay NTCLE100E3 Thermistor Datasheet: http://www.vishay.com/docs/29049/ntcle100.pdf

APPENDIX C: COMPONENT SYMBOL GLOSSARY

Passives

Resistor	Potentiometer	Photoresistor (LDR)

Capacitor	Polarized Capacitor	Variable Capacitor

Crystal	Inductor	Transformer

Diodes

Switching or Rectifying Diode (anode — cathode)

Alternate Symbol

ELECTRONICS LABORATORY MANUAL

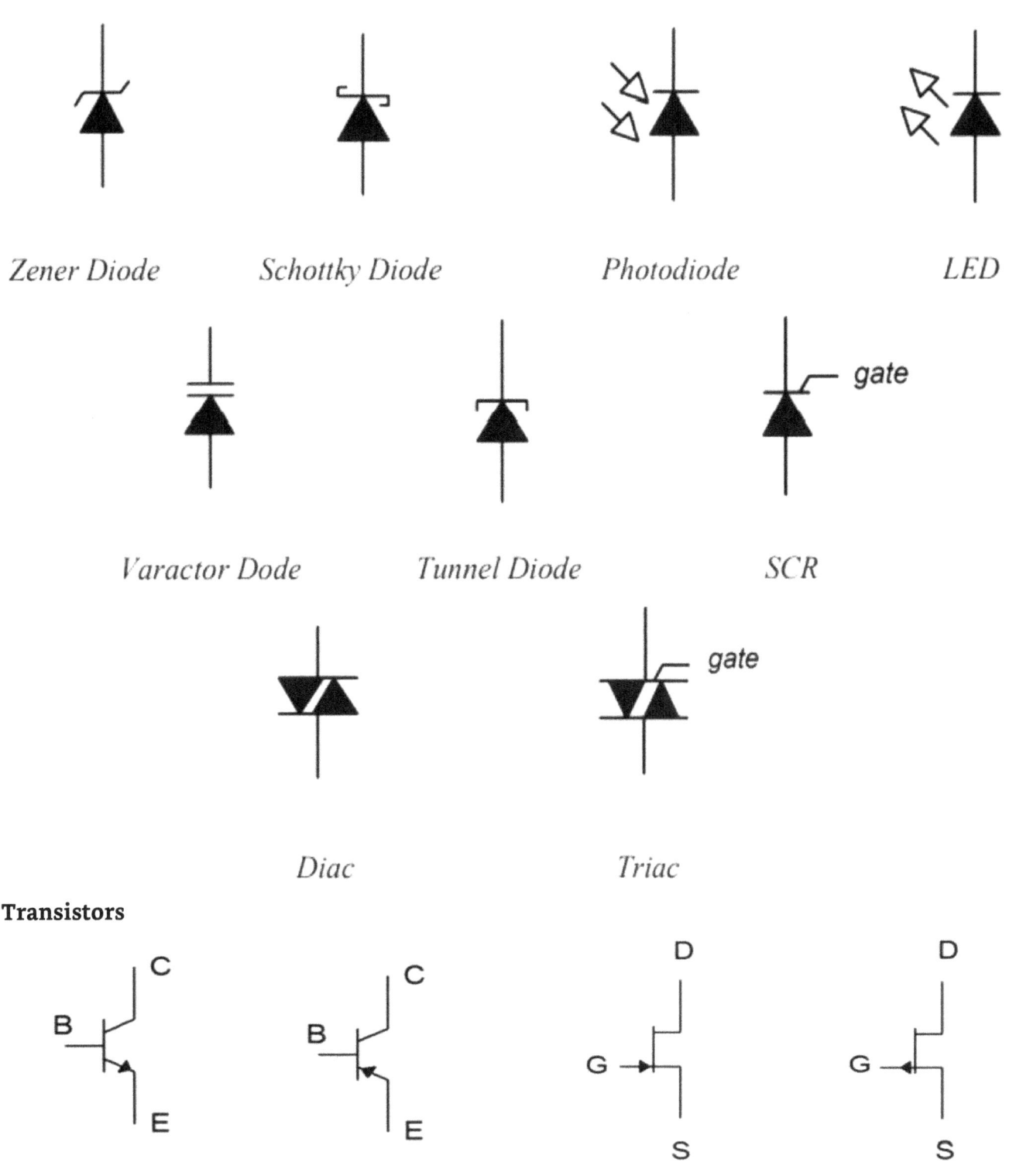

Transistors

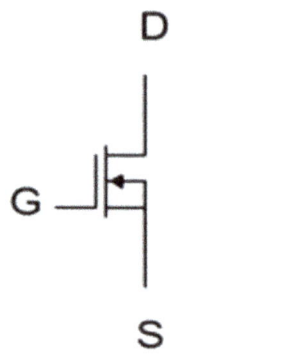 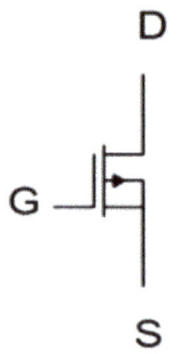 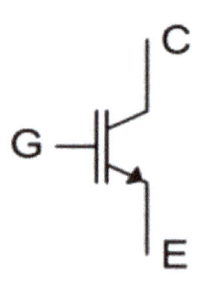

N Channel DE-MOSFET   *P Channel DE-MOSFET*   *IGBT*

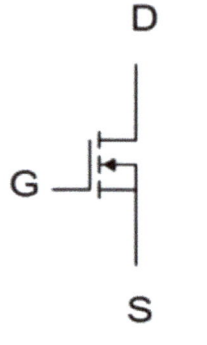 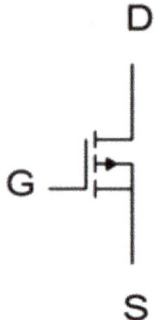 

N Channel E-MOSFET   *P Channel E-MOSFET*   *Phototransistor*

NB: *Transistors are sometimes drawn with a circle encompassing the body.*

REFERENCES

1. James M. Fiore, (2021), "Laboratory Manual for Semiconductor Devices: Theory and Application"
2. C. K. Alexander and M. Sadiku, *Fundamentals of Electric Circuits*, *4th Ed*,
3. https://en.wikipedia.org/wiki/Alternating_current Retrieved December, 2024
4. https://study.com/academy/lesson/alternating-current-definition-advantages-disadvantages.html
 Retrieved December, 2024

www.ingramcontent.com/pod-product-compliance
Lightning Source LLC
Chambersburg PA
CBHW062105220526
45471CB00010B/3601